SpringerBriefs in Computer Science

Series editors

Stan Zdonik, Brown University, Providence, RI, USA
Shashi Shekhar, University of Minnesota, Minneapolis, MN, USA
Xindong Wu, University of Vermont, Burlington, VT, USA
Lakhmi C. Jain, University of South Australia, Adelaide, SA, Australia
David Padua, University of Illinois Urbana-Champaign, Urbana, IL, USA
Xuemin Sherman Shen, University of Waterloo, Waterloo, ON, Canada
Borko Furht, Florida Atlantic University, Boca Raton, FL, USA
V. S. Subrahmanian, University of Maryland, College Park, MD, USA
Martial Hebert, Carnegie Mellon University, Pittsburgh, PA, USA
Katsushi Ikeuchi, University of Tokyo, Tokyo, Japan
Bruno Siciliano, Università di Napoli Federico II, Napoli, Italy
Sushil Jajodia, George Mason University, Fairfax, VA, USA
Newton Lee, Institute for Education, Research, and Scholarships, Los Angeles, CA, USA

W0225846

SpringerBriefs present concise summaries of cutting-edge research and practical applications across a wide spectrum of fields. Featuring compact volumes of 50 to 125 pages, the series covers a range of content from professional to academic.

Typical topics might include:

- A timely report of state-of-the art analytical techniques
- A bridge between new research results, as published in journal articles, and a contextual literature review
- A snapshot of a hot or emerging topic
- An in-depth case study or clinical example
- A presentation of core concepts that students must understand in order to make independent contributions

Briefs allow authors to present their ideas and readers to absorb them with minimal time investment. Briefs will be published as part of Springer's eBook collection, with millions of users worldwide. In addition, Briefs will be available for individual print and electronic purchase. Briefs are characterized by fast, global electronic dissemination, standard publishing contracts, easy-to-use manuscript preparation and formatting guidelines, and expedited production schedules. We aim for publication 8–12 weeks after acceptance. Both solicited and unsolicited manuscripts are considered for publication in this series.

More information about this series at http://www.springer.com/series/10028

Serghei Mangul · Harry Taegyun Yang ·
Eleazar Eskin · Noah Zaitlen

Hidden Treasures in Contemporary RNA Sequencing

 Springer

Serghei Mangul
Department of Computer Science
Institute for Quantitative
and Computational Biosciences
University of California Los Angeles
Los Angeles, CA, USA

Eleazar Eskin
Department of Computer Science
Department of Human Genetics
University of California Los Angeles
Los Angeles, CA, USA

Harry Taegyun Yang
Department of Computer Science
University of California Los Angeles
Los Angeles, CA, USA

Noah Zaitlen
Division of Pulmonary, Critical Care
Sleep and Allergy, Department of Medicine
Cardiovascular Research Institute
University of California
San Francisco, CA, USA

ISSN 2191-5768   ISSN 2191-5776   (electronic)
SpringerBriefs in Computer Science
ISBN 978-3-030-13972-8   ISBN 978-3-030-13973-5   (eBook)
https://doi.org/10.1007/978-3-030-13973-5

Library of Congress Control Number: 2019932600

This Springer imprint is published by the registered company Springer Nature Switzerland AG
The registered company address is: Gewerbestrasse 11, 6330 Cham, Switzerland

Contents

Hidden Treasures in Contemporary RNA Sequencing

Abstract High throughput RNA sequencing technologies have provided unprecedented opportunity to explore the individual transcriptome. Unmapped reads, the reads falling to map to the human reference, are a large and often overlooked output of standard RNA-Seq analyses; the hidden treasure in the contemporary RNA-Seq analysis is within the unmapped reads, illuminating previously unexplored biological insights. Here we develop Read Origin Protocol (ROP) to discover the source of all reads originating from complex RNA molecules, recombinant T and B cell receptors, and microbial communities. We applied ROP to 10,641 samples across 2630 individuals from 54 diverse adult human tissues. Our approach can account for 99.9% of 1 trillion reads of various read length. Using in-house RNA-Seq data, we show that immune profiles of asthmatic individuals are significantly different from the profiles of control individuals, with decreased average per sample T and B cell receptor diversity. We also show that microbiomes can be detected in human bloods via RNA-Sequencing and may elucidate important clinical changes in patients with schizophrenia. Furthermore, we demonstrate that receptor-derived reads among other hidden reads can be used to characterize the overall Ig repertoire across diverse human tissues using RNA-Sequencing. Our results demonstrate the potential of ROP to exploit the hidden treasure in contemporary RNA-Sequencing in order to better understand the functional mechanisms underlying connections between the immune system, microbiome, human gene expression, and disease etiology.

Keywords RNA Sequencing · B and T cell receptor immune repertoires · Human Microbiome · Unmapped reads · Immune system · RNA aligners

Main Text

Advances in RNA sequencing (RNA-Seq) technology have provided an unprecedented opportunity to explore gene expression across individual, tissues, and environments (Cloonan et al., 2008; Sultan et al., 2014; Tang et al., 2009) by efficiently

Serghei Mangul, Harry Taegyun Yang, Eleazar Eskin and Noah Zaitlen—These authors equally contributed in this work.

profiling the RNA sequences present in a sample of interest (Wang, Gerstein, & Snyder, 2009). RNA-Seq experiments currently produce tens of millions of short read subsequences sampled from the complete set of RNA transcripts that are provided to the sequencing platform. An increasing number of bioinformatic protocols are being developed to analyze reads in order to annotate and quantify the sample's transcriptome (Mihaela Pertea, 2015; Nicolae, Mangul, Mandoiu, & Zelikovsky, 2011; Trapnell et al., 2010). When a reference genome sequence or, preferably, a transcriptome of the sample is available, mapping-based RNA-Seq analysis protocols align the RNA-Seq reads to the reference sequences, identify novel transcripts, and quantify the abundance of expressed transcripts.

Unmapped reads, the reads falling to map to the human reference, are a large and often overlooked output of standard RNA-Seq analyses. Even in carefully executed experiments, the *unmapped reads* can comprise a substantial fraction of the complete set of reads produced; for example, approximately 9–20% of reads are unmapped in recent large human RNA-Seq projects (GTEx Consortium 2015; Li et al., 2014; Seqc/Maqc-Iii Consortium, 2014). Unmapped reads can arise due to technical sequencing artifacts that were produced by low quality and error prone copies of the nascent RNA sequence being sampled (Ozsolak & Milos, 2011). A recent study by Baruzzo et al. (2016) suggests that at least 10% of the reads simulated from human references remain unmapped across 14 contemporary state-of-the art RNA alligners. This rate may be due to shortcomings of the aligner's efficient yet heuristic algorithms (Siragusa, Weese, & Reinert, 2013). Reads can also remain unmapped due to unknown transcripts (Grabherr et al., 2011), recombined B and T cell receptor sequences (Blachly et al., 2015; Strauli & Hernandez, 2016), A-to-G mismatches from A-to-I RNA editing (Porath, Carmi, & Levanon, 2014), trans-splicing (Wu, Yi, Zhang, Zhou, & Sun, 2014), gene fusion (Wang et al., 2009), circular RNAs (Jeck & Sharpless, 2014), and the presence of non-host RNA sequences (Kostic et al., 2011) (e.g., bacterial, fungal, and viral organisms). By investigating the origin of unmapped reads, hidden biological insights can be discovered beyond the transcriptome characterized by conventional RNA-Seq analysis methods.

In this work, we report the development of a comprehensive method that can characterize the origin of unmapped reads obtained by RNA-seq experiments. Analyzing unmapped reads can inform future development of read mapping methods, provide access to additional biological information, and resolve the irksome puzzle of the origin of unmapped reads. We developed the Read Origin Protocol (ROP), a multi-step approach that leverages accurate alignment methods for both host and microbial sequences. The ROP tool contains a combination of novel algorithms and existing tools focused on specific categories of *unmapped reads* (Blachly et al., 2015; Brown, Raeburn, & Holt, 2015; Chuang et al., 2015; Kostic et al., 2011; Strauli & Hernandez, 2015). The comprehensive analytic nature of the ROP tool prevents biases that can otherwise arise when using standard targeted analyses. Currently, ROP supports human and mouse RNA-Seq data. ROP offers flexible interface to customization the computational tools used in the protocol.

Mapping-based RNA-Seq analysis protocols overlook reads that fail to map onto the human reference sequences (i.e., *unmapped reads*). We designed a read origin

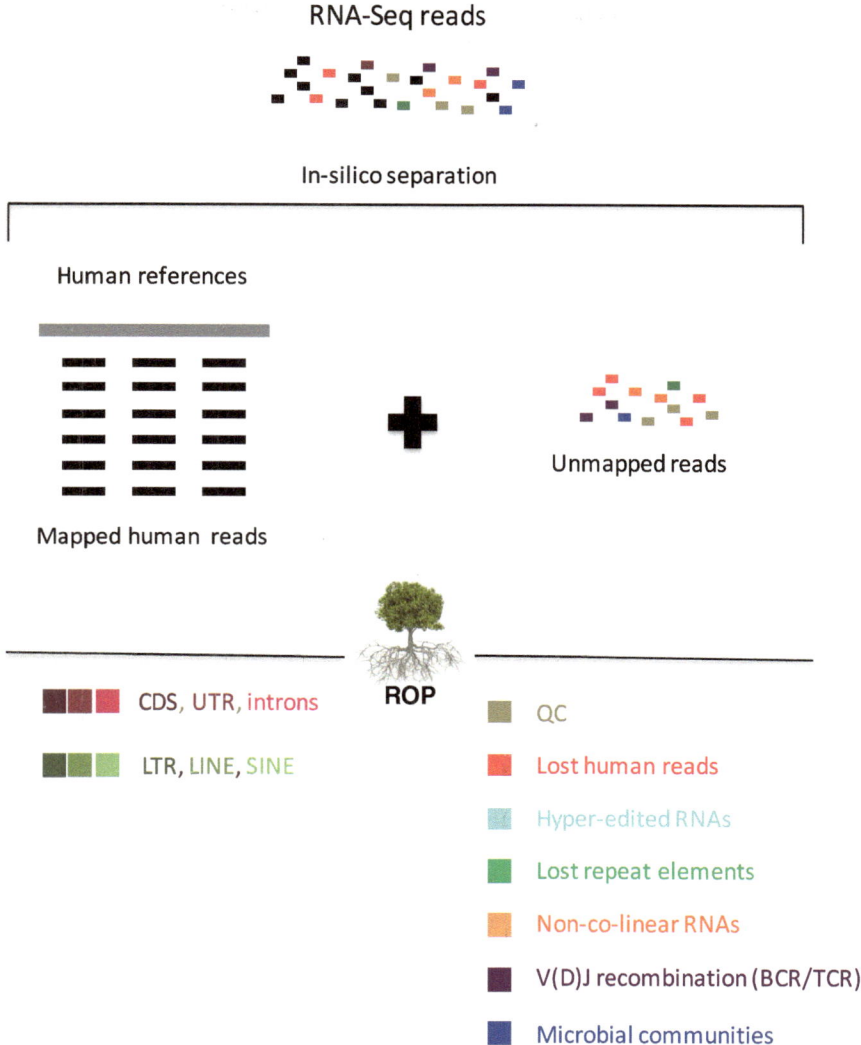

Fig. 1 Schematic of the Read Origin Protocol (ROP). Human reads are identified by mapping all reads onto the reference sequences using a standard high-throughput mapping algorithm. ROP protocol categorizes mapped reads into genomic (red colors) and repetitive (green colors) categories. Unmapped reads that fail to map are extracted and further filtered to exclude low quality reads, low complexity reads, and reads from rRNA repeats (brown color). ROP protocol is able to identify unmapped reads aligned to human references with use of a more sensitive alignment tool (lost human reads: red color), unmapped reads aligned to human references with excessive ('hyper') editing (hyper-edited RNAs: cyan color), unmapped reads aligned to the repeat sequences (lost repeat elements: green color), unmapped reads spanning sequences from distant loci (non-co-linear: orange color), unmapped reads spanning antigen receptor gene rearrangement in the variable domain (V(D)J recombination of BCR and TCR: violet color), and unmapped reads aligned to the microbial reference genomes and marker genes (microbial reads: blue color)

protocol (ROP) that identifies the origin of both mapped and unmapped reads (Fig. 1). The protocol first identifies human reads by mapping them onto a reference genome and transcriptome using a standard high-throughput mapping algorithm (Kim et al., 2013). We used tophat v. 2.0.12 with ENSEMBL GRCh37 transcriptome and hg19 build, but many other mapping tools are available and have recently been reviewed by Baruzzo et al. 2016. After alignment, reads are grouped into genomic (e.g., CDS, UTRs, introns) and repetitive (e.g., SINEs, LINEs, LTRs) categories. The rest of the ROP protocol characterizes the remaining *unmapped reads*, which failed to map to the human reference sequences.

To distinguish the mapped reads and unmapped reads, we mapped reads onto the human transcriptome (Ensembl GRCh37) and genome reference (Ensembl hg19) using TopHat2 (Kim et al., 2013) with the default parameters. Tophat2 was supplied with a set of known transcripts to optimize the transcriptomic mapping. ROP categorizes the mapped reads into genomic categories (junction read, CDS, intron, UTR3, UTR5, introns, inter-genic read, deep a deep inter-genic read, mitochondrial read, and multi-mapped read) based on their compatibility with the features defined by Ensembl (GRCh37) gene annotations. Then, ROP categorizes reads into repeat elements (classes and families) based on their compatibility with repeat instances defined by RepeatMasker annotation (Repeatmasker v3.3, Repeat Library 20120124) (Tarailo-Graovac & Chen, 2009). We count the number of reads overlapping variable(V), diversity (D), joining (J), and constant (C) gene segments of B cell receptor (BCR) and T cell receptor (TCR) loci using htseq-count (HTSeq v0.6.1) (Anders, Pyl, & Huber, 2014).

After the alignment and categorization of mapped reads, ROP processes unmapped reads to identify the origin of those reads. The ROP protocol effectively processes the unmapped reads in seven steps. The pairing information of the unmapped reads is disregarded, and each read from the pair is counted separately. First, we apply a quality control step to exclude low-quality reads, low-complexity reads, and reads that match rRNA repeat units among the unmapped reads [FASTQC (Andrews, 2010), SEQCLEAN ("https://sourceforge.net/projects/seqclean/," n.d.)]. Next, we employ Megablast (Camacho et al., 2009), a more sensitive alignment method, to search for human reads missed due to heuristics implemented for computational speed in conventional aligners and reads with additional mismatches. These reads typically include those with mismatches and short gaps relative to the reference set, but they can also include perfectly matched reads. Hyper-editing pipelines recognize reads with excessive ('hyper') editing, which are usually rejected by standard alignment methods due to many A-to-G mismatches (Porath et al., 2014). We use a database of repeat sequences to identify lost repeat reads among the *unmapped reads*. Megablast, and similar sensitive alignment methods, are not designed to identify 'non-co-linear' RNA (Chuang et al., 2015) reads from circRNAs, gene fusions, and trans-splicing events, which combine a sequence from distant elements. For this task, we independently map 20 bp read anchors onto the genome (see Supplemental Methods). Similarly, reads from BCR and TCR loci, which are subject to recombination and somatic hyper-mutation (SHM), require specifically designed methods. For this case, we use IgBlast (Ye, Ma, Madden, & Ostell, 2013). The remaining reads that did not map to any known human sequence are potentially microbial in origin. We use microbial genomes and phylogenetic marker genes to identify microbial reads and assign them to corresponding taxa (Truong et al., 2015). Microbial reads can be introduced by contamination or natural microbiome content in the sample, such as viral, bacterial, fungi, or other microbial species (Salter et al., 2014).

Taken together, ROP considers six classes of *unmapped reads*: (1) lost human reads, (2) hyper-edited reads, (3) lost repeat elements, (4) reads from 'non-co-linear' (NCL) RNAs, (5) reads from the recombination of BCR and TCR segments (i.e. V(D)J recombination), and (6) microbial reads. Previously proposed individual methods do examine some of these classes (Blachly et al., 2015; Brown, Raeburn, & Holt, 2015; Chuang et al., 2015; Kostic et al., 2011; Strauli & Hernandez, 2015). However, we find that performing a sequential analysis, in the order described above, is critical for minimizing misclassification of reads due to homologous sequences between the different classes. Furthermore, only a comprehensive analysis allows comparison across these classes.

Prior to the applying ROP on real data, it is imperative to confirm the correctness of read assignment. To demonstrate the accuracy of ROP's read assignment, we simulated RNA-Seq data as a mixture of transcriptomic, repeat, immune and microbial reads. We first map the RNA-Seq reads using TopHat2 aligner. TopHat2 was able to map 75.1% of transcriptomic reads. In addition to transcriptomic reads, it mapped 59.9% of repeat reads; and 80.7% of immune reads (Table 1a). We consider categorizing repeat and immune reads as human reads by TopHat2 or lost human reads by ROP as correct assignment, due to the presence of repeat sequences and immune genes in the human genome. Running ROP on unmapped reads provided 23.4% of transcriptomic reads; 39.2% of repeat reads; 12.3% of immune reads; and 100% of microbial reads. Altering the order of steps executed by ROP analysis resulted in 3.0% of repeat reads and 0.01% of transcriptomic reads to be reclassified as microbial (Table 1b–d). Immune reads spanning V(D)J recombinations may not sufficiently overlap V and J genes for a reliable identification, and 7.7% of those reads were missed by the ROP protocol.

In additional to the simulated data, we have used TCRB-Seq data prepared from three samples of kidney renal clear cell carcinoma (KIRC) by Li, Bo, et al. to demonstrate the assignment accuracy of immune reads. We downloaded matching RNA-Seq samples from the TCGA portal. In total, we obtained 301 million reads from three RNA-Seq samples. We considered the recombinations of V and J genes obtained from TCRB-Seq as the total immune repertoire. On average, ROP is able to capture 4.3% of total immune repertoire. All T cell receptor recombinations detected by ROP were confirmed by TCRB-Seq.

As it is demonstrated ROP can successfully assign the origin of the reads, ROP was applied to RNA-Seq samples to investigate the read origins in real samples. We applied ROP to one trillion RNA-Seq reads across 54 tissues from 2630 individuals. The data was combined from 3 studies: (1) in-house RNA-Seq data (n = 86) from the peripheral blood, nasal, and large airway epithelium of asthmatic and control individuals (S1); (2) multi-tissue RNA-Seq data from Genotype-Tissue Expression (GTEx v6) from 53 human body sites (GTEx Consortium 2015) (n = 8555) (S2);

Table 1 The effect of altering order of ROP step on the classification accuracy

Order of steps		Transcriptomics reads	Repeat reads	BCRs/TCRs	Microbiome reads
(a)					
	Total number of reads	1226	500	500	500
1	TopHat2	933	99	307	0
2	ROP step "lost human reads"	288	109	116	0
3	ROP step "lost repeat reads"	0	283	0	0
4	ROP step "immune (BCRs/TCRs)"	0	0	7	0
5	ROP step "microbiome reads"	0	0	0	500
	Unaccounted	5	9	70	0
(b)					
	Total number of reads	1226	500	500	500
1	TopHat2	933	99	307	0
3	ROP step "lost human reads"	288	1	116	0
2	ROP step "lost repeat reads"	0	391	0	0
4	ROP step "immune (BCRs/TCRs)"	0	0	7	0
5	ROP step "microbiome reads"	0	0	0	500
	Unaccounted	5	9	70	0
(c)					
	Total number of reads	1226	500	500	500
1	TopHat2	933	99	307	0
3	ROP step "lost human reads"	288	109	116	0
4	ROP step "lost repeat reads"	0	283	0	0
2	ROP step "immune (BCRs/TCRs)"	0	0	7	0
5	ROP step "xmicrobiome reads"	0	0	0	500
	Unaccounted	5	9	70	0

(continued)

Table 1 (continued)

Order of steps		Transcriptomics reads	Repeat reads	BCRs/TCRs	Microbiome reads
(d)					
	Total number of reads	1226	500	500	500
1	TopHat2	933	99	307	0
3	ROP step "lost human reads"	287	79	116	0
4	ROP step "lost repeat reads"	0	283	0	0
5	ROP step "immune (BCRs/TCRs)"	0	0	7	0
2	ROP step "microbiome reads"	1	30	0	500
	Unaccounted	5	9	70	0

Each column represents a category from which reads were simulated. Each row represents a distribution of reads from the same class across ROP categories. The results are presented for both the default and the altered order of ROP steps. Reads were mapped with TopHat2. Order of steps is presented in column 1. (a) The default order of ROP steps: "lost human reads," "lost repeat reads," "immune reads," and "microbial reads." (b) Altered order of ROP steps: "lost repeat reads," lost human reads," "immune reads," and "microbial reads." (c) Altered order of ROP steps: "immune reads," lost human reads," "lost repeat reads," and "microbial reads." (d) Altered order of ROP steps: "microbial reads," "lost human reads," "lost repeat reads," "immune reads." Percentages are calculated from total number of reads from each category. We used reference human transcript sequences to simulate transcriptomics reads. We used reference repeat sequences to simulate repeat reads. Immune transcripts were simulated as recombinations of V D and J segments with non-template insertions at the junction points. We have use microbiome sequences downloaded from NCBI to simulate the microbial reads

(3) randomly selected RNA-Seq samples from the Sequence Read Archive (SRA) (n = 2000) (S3). Unless otherwise noted, we reported percentage of reads averaged across 3 datasets.

RNA-Seq data obtained from the three sources represent a large collection of tissue types and read diversity. We selected these three sources to most accurately model the precision and broad applicability of ROP. The in-house RNA-Seq data was collected from 53 asthmatics and 33 controls. RNA-Seq libraries were prepared from total RNA with two types of RNA enrichment methods: (1) Poly(A) enrichment libraries, applied to RNA from peripheral blood and nasal epithelium (n = 38), and (2) ribo-depletion libraries, applied to RNA from large airway epithelium (n = 49). The GTEx dataset was derived from 38 solid organ tissues, 11 brain sub-regions, whole blood, and three cell lines across 544 individuals. Randomly selected SRA RNA-Seq samples included samples from whole blood, brain, various cell lines, muscle, and placenta. Length of reads from in-house data was 100 bp, read length in GTEx data was 76 bp, read length in SRA data ranged from 36 bp to 100 bp. In total, 1 trillion reads (97 Tbp) derived from 10,641 samples were available for ROP. For

Table 2 RNA-Seq datasets overview

Datasets	S1	S2	S3
Number of samples	87	8555	2000
Read length	100bp	76bp	25-100bp
Average number of reads per sample, (million reads)	88.8	54.6	90.2
Percentage of mapped reads (%)	83.8	88.2	77.2

In-house RNA-Seq data (n = 86) from the peripheral blood, nasal, and large airway epithelium of asthmatic and control individuals (S1); (2) multi-tissue RNA-Seq data from Genotype-Tissue Expression (GTEx v6) from 53 human body sites (GTEx Consortium 2015) (n = 8555) (S2); (3) randomly selected RNA-Seq samples from the Sequence Read Archive (SRA) (n = 2000) (S3). Unless otherwise noted, we reported percentage of reads averaged across 3 datasets. For counting purposes, the pairing information of the reads is disregarded, and each read from a pair is counted separately

counting purposes, the pairing information of the reads is disregarded, and each read from a pair is counted separately. The diversity of the biological source of RNA-Seq provides an opportunity to investigate the impact of ROP on elucidating underlying information hidden within the unmapped reads (Table 2).

We used standard read mapping procedures to obtain mapped and unmapped reads from all three data sources. Read mapping for GTEx data was performed by the GTEx consortium using TopHat2 (Kim et al., 2013). Following the GTEx consortium practice, we used TopHat2 to map reads from in-house and SRA studies. ROP protocol allows user to map the reads with RNA-Seq aligner of choice. High-throughput mapping using TopHat2 (Kim et al., 2013) recovered 83.1% of all reads from three studies (Fig. 2a), with the smallest fraction of reads mapped in the SRA study (79% mapped reads). From the *unmapped reads*, we first excluded low-quality/low-complexity reads and reads mapping to the rRNA repeating unit, which together accounted for 7.0 and 2.4% of all reads, respectively (Fig. 2b). We were then able to align unmapped reads to human reference sequences (5.7% of all reads, Fig. 2c) and identify "hyper-edited" reads (0.1% of all reads Fig. 2d). We then referenced repeat sequences (0.2% of all reads, Fig. 2d), reads identified as 'non-co-linear'(NCL) RNAs (circRNAs, gene fusion or trans-splicing) (0.3% of all reads, Fig. 2e), and reads mapped to recombined B and T cell receptors (0.02% off all reads, Fig. 2f). The remaining reads were mapped to the microbial sequences (1.4% off all reads, Fig. 2g). Following the seven steps of ROP, the origins of 99.9% of reads were identified. Genomic profile of unmapped reads for each dataset is separately reported in Table 3. This resource allows the bioinformatics community to further increase the number of reads with known origin. By applying ROP, unmapped reads, commonly regarded as unusable waste reads, can be leveraged to illuminate the hidden treasure in RNA-Seq datasets.

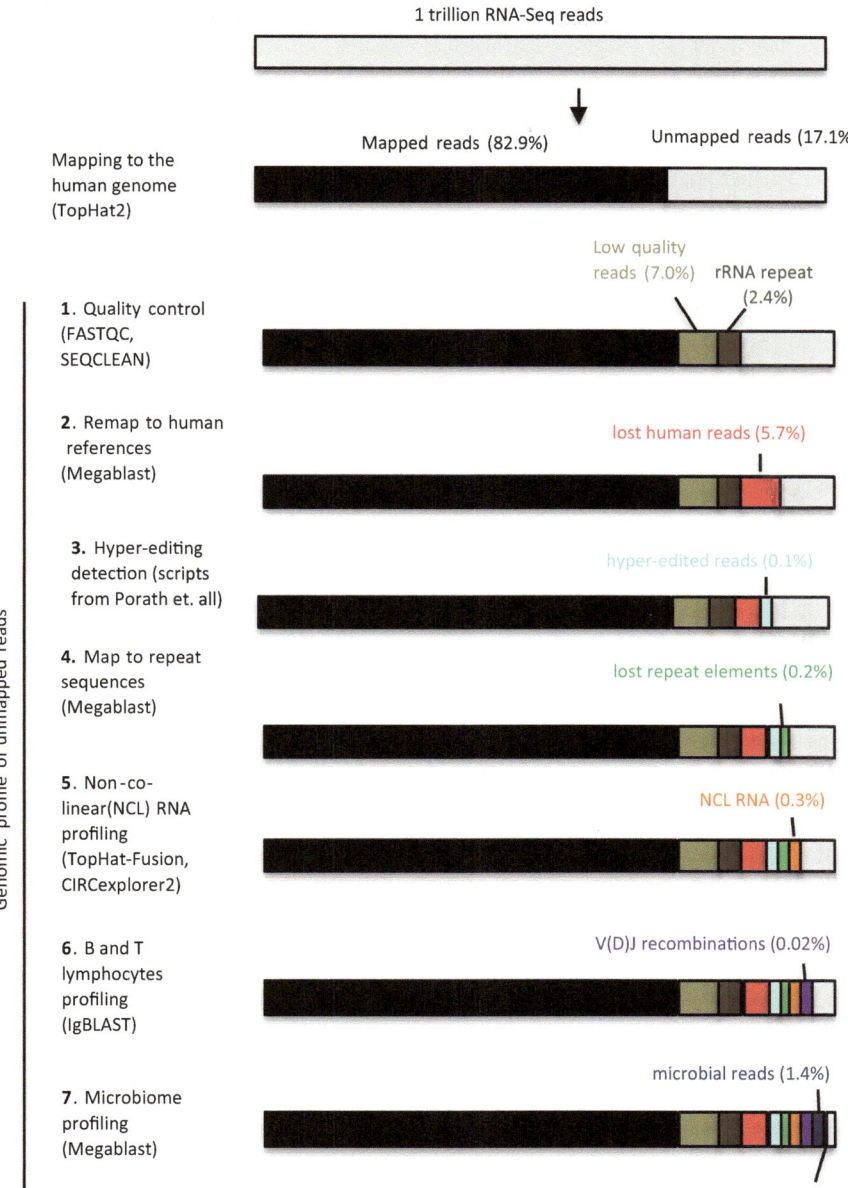

1 trillion RNA-Seq reads

Mapping to the human genome (TopHat2)

Mapped reads (82.9%)   Unmapped reads (17.1%)

1. Quality control (FASTQC, SEQCLEAN)

Low quality reads (7.0%)   rRNA repeat (2.4%)

2. Remap to human references (Megablast)

lost human reads (5.7%)

3. Hyper-editing detection (scripts from Porath et. all)

hyper-edited reads (0.1%)

4. Map to repeat sequences (Megablast)

lost repeat elements (0.2%)

5. Non-co-linear(NCL) RNA profiling (TopHat-Fusion, CIRCexplorer2)

NCL RNA (0.3%)

6. B and T lymphocytes profiling (IgBLAST)

V(D)J recombinations (0.02%)

7. Microbiome profiling (Megablast)

microbial reads (1.4%)

unaccounted reads (0.1%)

Genomic profile of unmapped reads

◄**Fig. 2 Genomic profile of unmapped reads across 10,641 samples and 54 tissues**. Percentage of unmapped reads for each category is calculated as a fraction from the total number of reads. Bars of the plot are not scaled. Human reads (black color) mapped to reference genome and transcriptome via TopHat2. Unmapped reads are profiled using seven steps of ROP protocol, described below. (1) Low quality/low-complexity (light brown) and reads matching rRNA repeating unit (dark brown) were excluded. (2) ROP identifies lost human reads (red color) from unmapped reads using a more sensitive alignment. (3) Hyper-edited reads are captured by hyper-editing pipeline proposed in (Porath et al., 2014). (4) ROP identifies lost repeat sequences (green color) by mapping unmapped reads onto the reference repeat sequences. (5) Reads arising from trans-spicing, gene fusion and circRNA events (orange color) are captured by a TopHat-Fusion and CIRCexplorer2 tools. (6) IgBlast is used to identify reads spanning B and T cell receptor gene rearrangement in the variable domain (V(D)J recombinations) (violet color). (7) Microbial reads (blue color) are captured by mapping the reads onto the microbial reference genomes

Table 3   Genomic profile of unmapped reads reported for each dataset (S1, S2, S3)

	S1	S2	S3	Averaged across 3 datasets
Mapped	83.2%	88.2%	77.2%	82.9%
Unmapped	17%	11.8%	23%	17.1%
Low quality reads	4.8%	7.0%	9%	7.0%
rRNA repeat	3.8%	0.1%	3%	2.4%
Lost human reads	6.0%	3.7%	8%	5.7%
Hyper-edited reads	0.02%	0.1%	0.1%	0.1%
Lost repeat reads	0.3%	0.1%	0.1%	0.2%
NCL RNA	0.3%	0.3%	0.4%	0.3%
V(D)J recombinations	0.01%	0.03%	0.01%	0.02%
Microbial reads	1.5%	0.5%	2.3%	1.4%
Unaccounted reads	0.18%	0.09%	0.10%	0.12%

Percentage of unmapped reads for each category is calculated as a fraction from the total number of reads. Bars of the plot are not scaled. Human reads (black color) mapped to reference genome and transcriptome via TopHat2. (a) Low quality/low-complexity (light brown) and reads matching rRNA repeating unit (dark brown) were excluded. (b) Hyper-edited reads are captured by hyper-editing pipeline proposed in (Porath et al., 2014). (c) ROP identifies lost human reads (red color) from unmapped reads using a more sensitive alignment. (d) ROP identifies lost repeat sequences (green color) by mapping unmapped reads onto the reference repeat sequences. (e) Reads arising from trans-spicing, gene fusion and circRNA events (orange color) are captured by a TopHat-Fusion and CIRCexplorer2 tools. (f) IgBlast is used to identify reads spanning B and T cell receptor gene rearrangement in the variable domain (V(D)J recombinations) (violet color). (g) Microbial reads (blue color) are captured by mapping the reads onto the microbial reference genomes

Table 4 Average mapping rate for different aligners with different mapping settings

	Default/fast	Sensitive	Very sensitive
Tophat2	89.06% (3.84)	89.22% (3.51)	89.18% (3.62)
STAR	80.86% (9.22)	81.70% (9.25)	81.74% (9.35)

The average rate is noted, and the standard deviation is noted within parenthesis

Although most of reads from human RNA are aligned to human genome and transcriptome, some human reads may remain unmapped due to the heuristic nature of high throughput aligners (Baruzzo et al., 2016; Siragusa et al., 2013). As shown by Baruzzo et al., even the best performing RNA-Seq aligners fail to map at least 10% of reads simulated from the human references. To prevent misclassification of reads derived from human genome into other downstream ROP categories, we used the slower and more sensitive Megablast aligner on this subset of unmapped reads. This method allows us to filter an additional 5.7% of human reads.

To minimize the number of unmapped human reads in the mapping step, we investigated the impact of mapping parameters and RNA-Seq aligners on the number of unmapped reads. We additionally used STAR (Dobin et al., 2013) and added results for sensitive and very sensitive mapping settings of each of the tools. We observe that an alternative aligner and a more sensitive mapping setting has no substantial effect on the number of mapped reads (Table 4). This is in line with Baruzzo et al., 2016, which have shown that optimizing the parameters of RNA-Seq aligner is a non-trivial task and methods with good performance for the default setting is a preferred choice. Nevertheless, the impact of aligner choices is tested on other commonly used contemporary RNA-Seq to ensure the robustness of the premise of ROP, mining the hidden treasure in the uncharted territory of unmapped reads.

Following Baruzzo et al. (2016), we have selected five contemporary RNA-Seq aligners that were reported to map a minimum of 75% of the reads simulated from human transcriptome using recommended parameters, HISAT2, Novoalign, TopHat2, and STAR. We used the same simulated data, explained above, comprising transcriptomic, repeat, immune, and microbial reads, to investigate the effect of RNA-Seq aligner on the number of mapped reads. On average, 90.9% of simulated transcriptomic reads are mapped by RNA-Seq aligners with default and tuned parameters. STAR (default and tuned setting) and Novoalign (tuned settings) are able to map >99% of transcriptomic reads. For other categories of simulated reads, RNA-Seq aligners were able to map 47.6% of reads (Fig. 3a).

We have investigated the effects that choice of RNA-Seq aligner can have on the fraction of reads accounted by ROP. Overall, ROP increases the number of categorized reads by an average of 39.5% across different RNA-Seq aligners. The best results were achieved by ROP using STAR (default and tuned setting), Novoalign (default and tuned settings), and HISAT2 (tuned setting), which allows ROP to account for >99.0% of all reads from RNA-Seq mixture (Fig. 3b). Both STAR and Novoalign perform extensive soft clipping (partial mapping), resulting in alignment of recombined V(D)J sequences, which are not part of the human genome (Fig. 4).

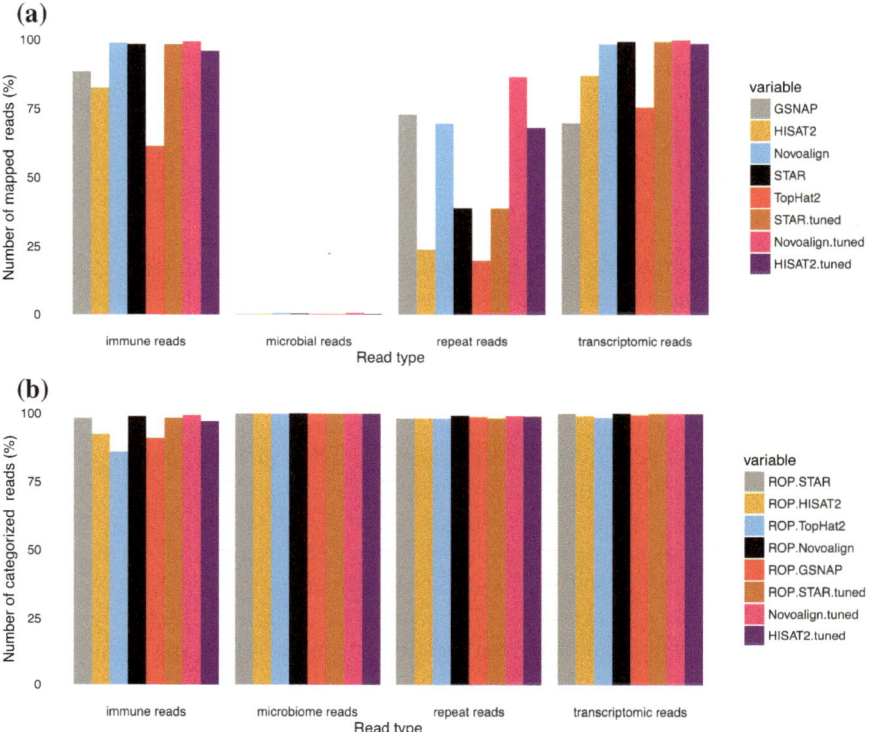

Fig. 3 The effect of RNA-Seq aligner on the fraction of reads accounted by ROP. Percentages are calculated from the total number of reads in each category. RNA-Seq aligners were run with default and optimized (tuned) parameters. We use tuned setting recommended by Baruzzo et al. (2016). TopHat2 and GSNAP were only run with default settings. Results are presented for simulated RNA-Seq data composed of transcriptomic, repeat, immune, and microbial reads. **a** Percentages of reads accounted by RNA-Seq aligners. **b** Percentages of reads categorized by ROP across five state of the art aligners

This can partially explain the increased number of mapped reads produced by ROP when compared to the results from other tools. Notably, Novoalign with default and tuned settings mapped 0.6% of microbial reads to the human genome, which corresponds to false-positive hits (Fig. 3b).

In addition to simulated data, we randomly selected ten SRA samples to investigate the effect of RNA-Seq aligner on the number of mapped reads from the real datasets. On average we observe 91.8% of reads mapped; the best performance is achieved by HISAT2 and STAR, which allow mapping of 93.1% and 92.7% of the reads, respectively (Fig. 5). Based on the achieved results and reported results in Baruzzo et al., we recommend using HISAT2 and STAR aligners to map the RNA-Seq reads and prepare unmapped reads. Novoalign is not recommended for use with ROP protocol due to its substantially longer running time, which makes it computationally infeasible. Thus, the choice of aligner does not have a substantial impact on mapping,

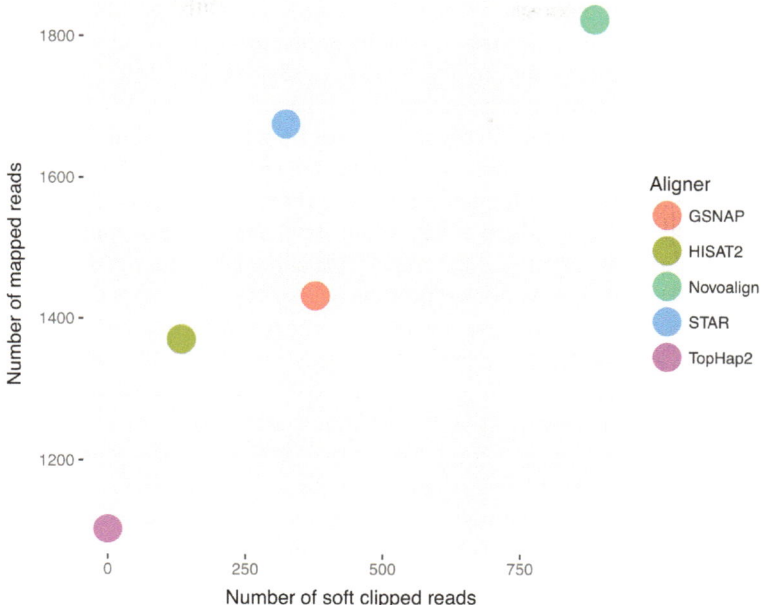

Fig. 4 Relationship between the number of soft clipped RNA-Seq reads (partially mapped reads) and the total number of reads. Results are presented for simulated RNA-Seq data composed of transcriptomic, repeat, immune, and microbial reads

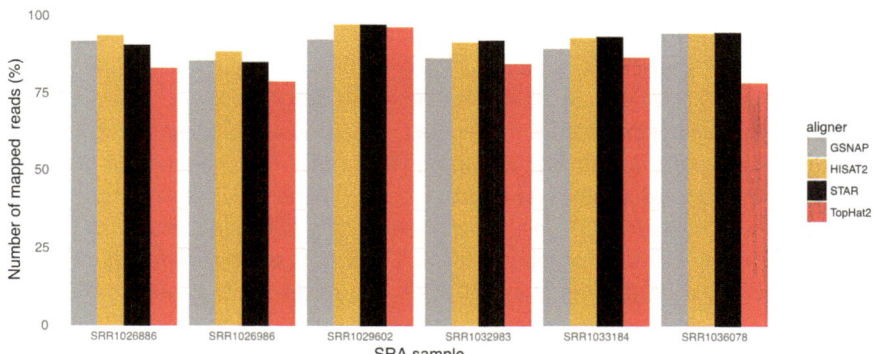

Fig. 5 Number of the RNA-Seq reads mapped to the human reference genome across five state-of-the-art RNA-Seq aligners. Number of mapped reads is separately reported for each SRA sample. Percentages are calculated from the total number of reads. Results are presented for 10 randomly selected SRA RNA-Seq samples. Tools were run with default parameters. Novoalign was excluded from this analysis because none of the experiments finished running within 24 h

and Read Origin Protocol can be applied to conventionally processed RNA-Seq samples for mining the hidden treasure in the unmapped portions of samples.

As the idea of mining the hidden treasure is demonstrated, it is necessary to explore the type of hidden treasures that can be unearthed to illuminate underlying biological information. Before investigating reads from foreign origin, it is imperative to identify reads from possible human origin that may not be aligned by contemporary alignment framework using human reference genome and transcriptome, such as repeat elements or hyper-edited reads. Using both mapped and unmapped reads across the studies, we classified on average 7.5% of the RNA-Seq reads as repetitive sequences originated from various repeat classes and families (Fig. 6). We observe Alu elements to have 33% relative abundance, which was the highest among all the repeat classes. Among DNA repeats, hAT-Charlie was the most abundant element with 50% relative abundance (Fig. 7). Among SVA retrotransposons, SVA-D was the most abundant element with 50% relative abundance (Fig. 8). Consistent with repEnrich (Criscione, Zhang, Thompson, Sedivy, & Neretti, 2014), when using in-house data we observe the differences in proportions of L1 and Alu elements between poly(A) and ribo-depletion libraries. Among the repeat reads, poly(A) samples have the highest fraction of reads mapped to Alu elements, and ribo-depleted samples have the highest fraction mapped to L1 elements (Fig. 9). Among the GTEx tissues, testis showed significantly higher expression of SVA F retrotransposons compared to other GTEx tissues ($p = 2.46 \times 10^{-33}$) (Fig. 10). Furthermore, we observe high co-expression of *Alu* elements and L1 elements across GTEx tissues ($R^2 = 0.7615$) (Fig. 11).

Using the standard read mapping approaches, some human reads may remain unmapped due to "hyper editing." An extremely common post-transcriptional modification of RNA transcripts in human is A-to-I RNA editing (Bazak et al., 2014), which alters the RNA-Seq base reads, leading to difficulty in correct alignments. We define a read to be hyper-edited if the number of A-to-G mismatches exceeds 5% of its length. Adenosine deaminases acting on RNA (ADARs) proteins can modify a genetically encoded adenosine (A) into an inosine (I). Inosine is read by the cellular machinery as a guanosine (G), and, in turn, sequencing of inosine results in G where the corresponding DNA sequencing reads A. Current methods to detect A-to-I editing sites are based on the alignment of RNA-Seq reads to the genome to identify such A-to-G mismatches. Reads with excessive ('hyper') editing are usually rejected by standard alignment methods. In this case, many A-to-G mismatches obscure their genomic origin.

We have identified hyper-edited reads by using the pipeline proposed in (Porath et al., 2014). This hyper-editing pipeline transforms all As into Gs, in both the unmapped reads and the reference genome, and the pipeline realigns the transformed RNA-Seq reads and the transformed reference genome. The method then recovers original sequences and searches for dense clusters of A-to-G mismatches. A total of 201,676,069 hyper-edited reads were identified across all samples from the three studies. As a control for the detection, we calculated the prevalence of all 6 possible nucleotide substitutions and found that 79.9% (201,676,069/252,376,867) of the detected reads were A-to-G mismatches (Fig. 12). In comparison, the in-house

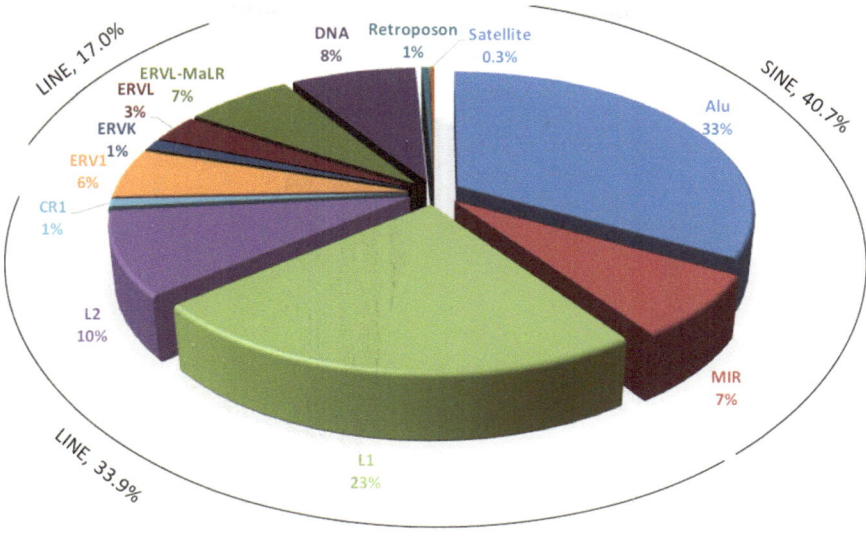

On average 7% of RNA-Seq reads are categorized as repeats

Fig. 6 Profile of repeat elements across based on repeat sequences inferred from mapped and unmapped reads (lost repeat reads). ROP identifies and categorizes repetitive sequences among the mapped and unmapped reads. Mapped reads were categorized based on the overlap with the repeat instances prepared from RepeatMasker annotation (Repeatmasker v3.3, Repeat Library 20120124). Lost repeat reads are unmapped RNA-Seq reads aligned onto the reference repeat sequences (prepared from Repbase v20.07). The percentages are the averages across 10,641 samples

RNA-Seq samples have a 96.1% rate of A-to-G mismatches. This massive over-representation of mismatches strongly suggests that these reads resulted from ADAR mediated RNA editing. However, additional experiments are required to confirm the nature of these edits. In addition, we found that the nucleotide sequence context of the detected editing sites complies with the typical sequence motif of ADAR targets (Fig. 13). The exclusion of these reads from pool allows ROP to investigate the non-conventional origins of the unmapped reads, meanwhile the careful process of classification of reads from conventional human origin can improve the alignment of missed human reads and elucidate the underlying prevalence of repeat elements and RNA editing process in human transcriptome.

After careful classification of unmapped reads through rigorous remapping process for conventional human transcriptome, ROP searches for complex transcriptomic events, such as non-colinearly spliced reads. The ROP protocol is able to detect 'non-co-linear' reads via Tophat-Fusion (Kim & Salzberg, 2011) and CIRC-explorer2 (Zhang et al., 2016) tools from three classes of events: (1) reads spliced distantly on the same chromosome supporting trans-splicing events; (2) reads spliced across different chromosomes supporting gene fusion events; and (3) reads spliced in a head-to-tail configuration supporting circRNAs. On average, we observed 816

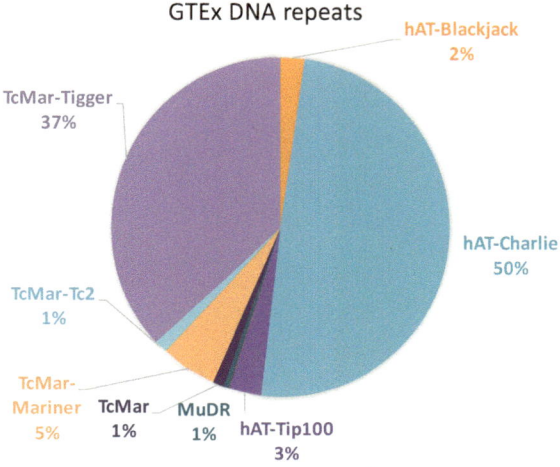

*Percentages are calculated as a fraction from the reads matching DNA repeats

Fig. 7 Profile of DNA repeats based on repeat sequences inferred from mapped and unmapped reads (lost repeat reads). ROP identifies and categorizes DNA repetitive sequences among the mapped and unmapped reads. Mapped reads were categorized based on the overlap with the repeat instances prepared from RepeatMasker annotation (Repeatmasker v3.3, Repeat Library 20120124). Lost repeat reads are unmapped RNA-Seq reads aligned onto the reference repeat sequences (prepared from Repbase v20.07). The percentages are the averages across 10,641 samples

Fig. 8 Profile of SVA retrotransposons based on repeat sequences inferred from mapped and unmapped reads (lost repeat reads). ROP identifies and categorizes SVA retrotransposons sequences among the mapped and unmapped reads. Mapped reads were categorized based on the overlap with the repeat instances prepared from RepeatMasker annotation (Repeatmasker v3.3, Repeat Library 20120124). Lost repeat reads are unmapped RNA-Seq reads aligned onto the reference repeat sequences (prepared from Repbase v20.07)

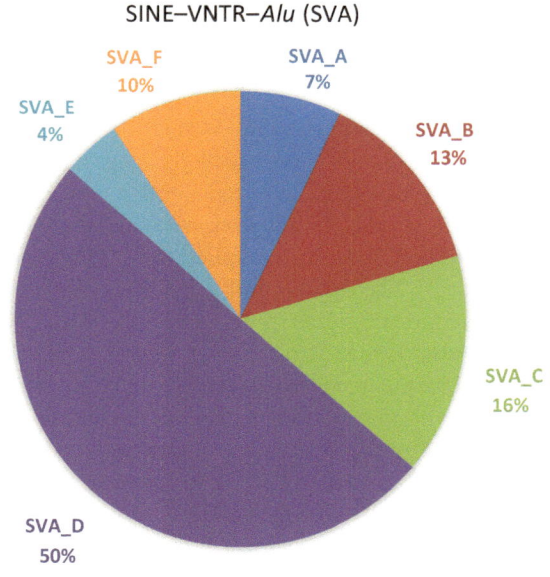

*Percentages are calculated as a fraction from the reads matching
SVA Retroposons

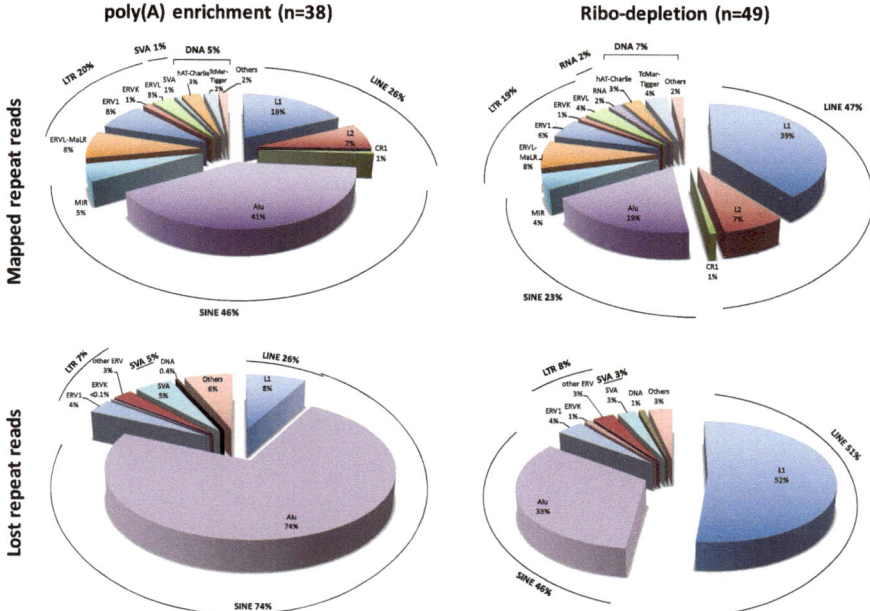

Fig. 9 Profile of repeat elements across poly(A) enrichment and ribo-depletion libraries. ROP identifies and categorizes repetitive sequences among the mapped and unmapped reads. RNA-Seq samples were prepared by poly(A) enrichment protocol (n = 38) and ribo-depletion protocol (n = 49). Mapped reads were categorized based on the overlap with the repeat instances prepared from RepeatMasker annotation (Repeatmasker v3.3, Repeat Library 20120124). Lost repeat reads are unmapped RNA-Seq reads aligned onto the reference repeat sequences (prepared from Repbase v20.07)

trans-splicing events, 7510 fusion events, and 930 circular events per individual sample supported by more than one read. Over 90% of non-co-linear events were supported by fewer than 10 samples (Fig. 14). We used a liberal threshold, based on number of reads and individuals, because our interest is mapping all reads. However, a more stringent cut off is recommended for confident identification of non-co-linear events, especially in the clinical settings.

Based on the in-house RNA-Seq data, we observe that the library preparation technique strongly affects the capture rate of non-co-linear transcripts. To compare the number of NCL events, we sub-sampled unmapped reads to 4,985,914 for each sample, which corresponded to the sample with the smallest number of unmapped reads among in-house RNA-Seq samples. We observed an average increase of 92% of circRNAs in samples prepared by ribo-depletion compared to poly(A) protocol (p-value = 3×10^{-12}) (Fig. 15). At the same time, we observed an average 43% decrease of trans-splicing and fusion events in samples prepared by ribo-depletion compared to poly(A) protocol (p-value < 8×10^{-4}) (Fig. 15). However, because the tissues differed between protocols (e.g., nasal versus large airway epithelium), this effect might be due in part to tissue differences in NCL events. We view the

Fig. 10 Average number of SVA-F reads across GTEx tissues. ROP identifies and categorizes *SVA* retrotransposons sequences among the mapped and unmapped reads. Mapped reads were categorized based on the overlap with the repeat instances prepared from RepeatMasker annotation (Repeatmasker v3.3, Repeat Library 20120124). Lost repeat reads are unmapped RNA-Seq reads aligned onto the reference repeat sequences (prepared from Repbase v20.07). Among the GTEx tissues, testis showed significantly higher expression of SVA F retrotransposons compared to other tissues ($p = 2.46 \times 10^{-33}$)

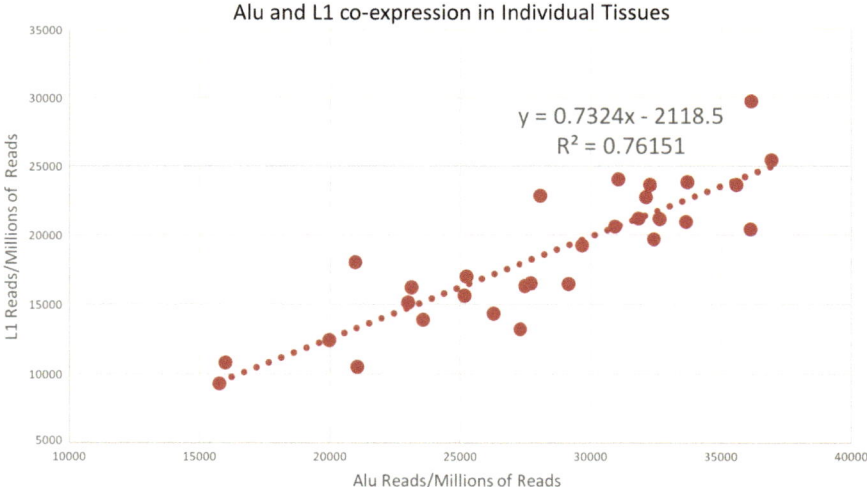

Fig. 11 Co-expression of Alu and L1 elements across GTEx tissues. ROP identifies and categorizes repetitive sequences among the mapped and unmapped reads. Mapped reads were categorized based on the overlap with the repeat instances prepared from RepeatMasker annotation (Repeatmasker v3.3, Repeat Library 20120124). Lost repeat reads are unmapped RNA-Seq reads aligned onto the reference repeat sequences (prepared from Repbase v20.07)

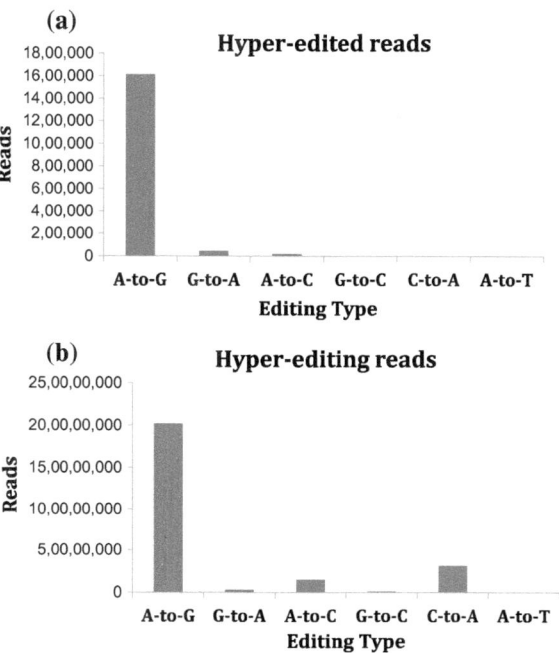

Fig. 12 Distribution of hyper-edited reads. a Hyper-editing identified in the in-house data. Results showed that 96% of the reads were A-to-G, indicating a high level of specificity for the hyper-editing screen. The 1,613,213 detected A-to-G reads contain 10,666,458 editing events (3,157,685 unique editing-sites). **b** Hyper-editing identified in the GTEx RNA-Seq data. Results showed that 80% of the reads were A-to-G, indicating a high level of specificity for the hyper-editing screen. The 201,676,069 detected A-to-G reads contain 1,130,591,911 editing events (690,386,562 unique editing-sites)

tissue differences effect to be unlikely. We previously showed that gene expression profiles of nasal airway tissue largely recapitulate expression profiles in the large airway epithelium tissue (Poole et al., 2014).

Furthermore, many NCL events will not be captured by poly-A selection. Therefore, we expect systematic differences in NCL abundance between capture methods. There were no statistically significant differences (p-value $> 5 \times 10^{-3}$) between NCL events in cases and controls. We have compared number of NCL reads across GTEx tissue, and we observe the highest fraction of NCL reads across pancreas samples with 0.75% of reads classified as NCL reads. In all other tissue types, ROP classified approximately 0.3% reads as NCL reads (Fig. 16).

After classifying NCL reads, ROP classified the remaining reads through human repertoire and microbial transcriptome databases. Reads mapped to B and T cell receptor loci and unmapped reads were used to survey the human adaptive immune repertoires in health and disease. We first used the mapped reads to extract reads entirely aligned to BCR and TCR genes. Using IgBlast (Ye et al., 2013), we identified unmapped reads with extensive somatic hyper mutations (SHM) and reads arising

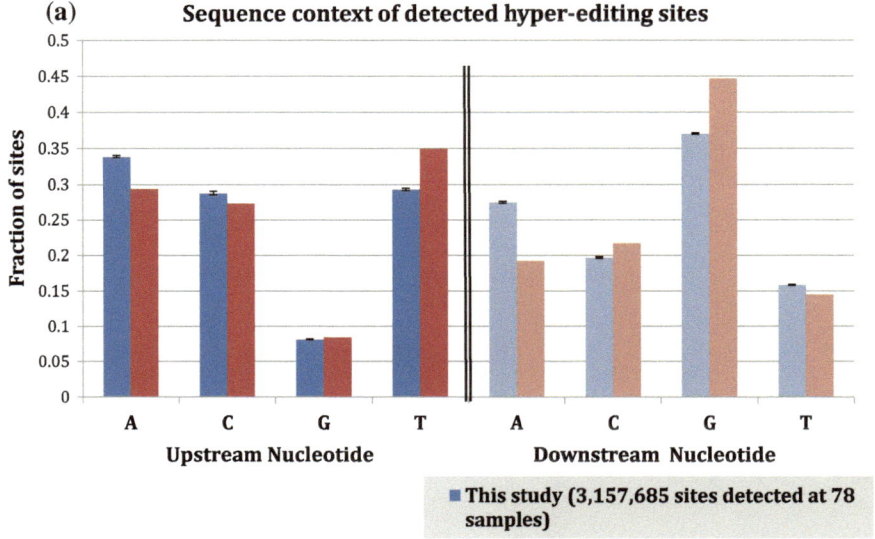

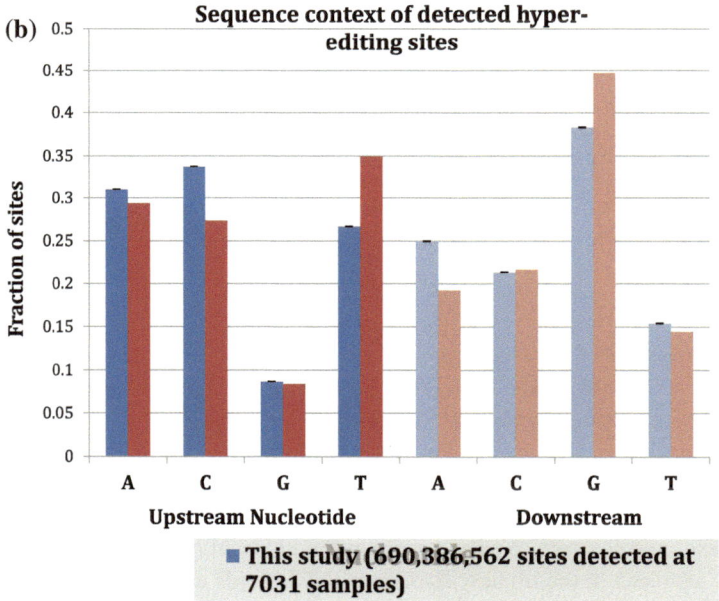

Fig. 13 The sequence context of the Figure S8. The sequence context of the detected hyper-edited A-to-G sites. The sequence near the detected hyper-editing sites is depleted of Gs upstream and enriched with Gs downstream, in agreement with previously known data about the ADAR motif. The bars correspond to the fraction of editing sites with each type of nucleotide one base upstream and downstream of the site. Results are shown for sites detected in-house RNA-Seq data (**a**) and GTEx RNA-Seq data (**b**) using the hyper-editing pipeline and human editing-sites from the RADAR database

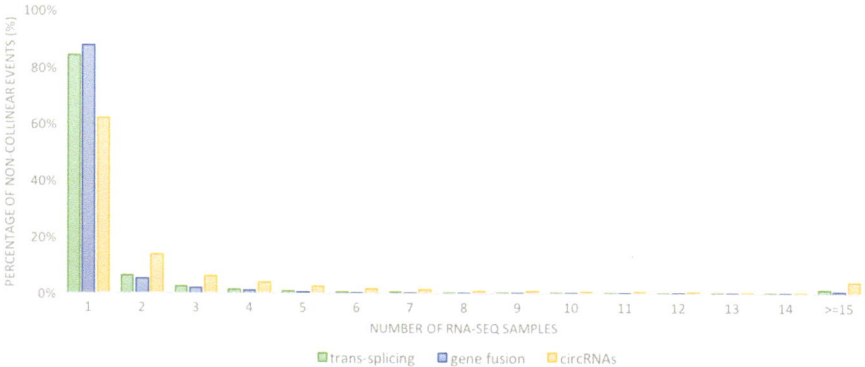

Fig. 14 Distribution of non-co-linear (NCL) events across across 10,641 samples. Reads arising from trans-splicing, gene fusion and circRNA events are captured by a TopHat-Fusion and CIRCex-plorer2 tools. Trans-splicing events are identified from reads that are spliced distantly on the same chromosome. Gene fusion events are identified from reads spliced across different chromosomes. CircRNAs are identified from reads spliced in a head-to-tail configuration

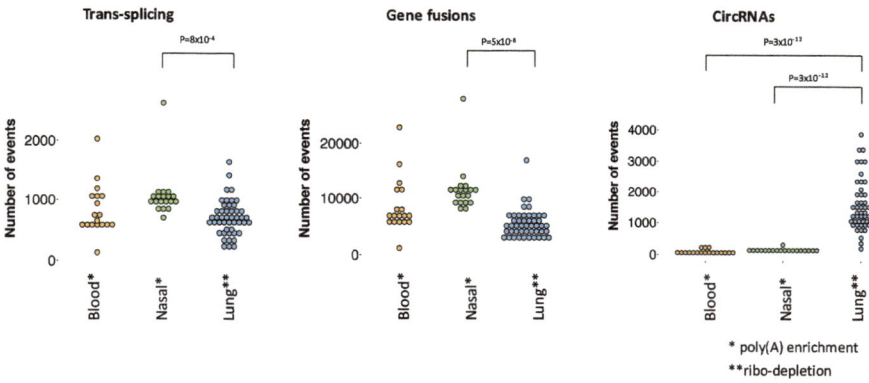

Fig. 15 Number of NCL events across in-house tissues and library preparation protocols. NCL events per sample are detected by TopHat-Fusion and CIRCexplorer tools. Samples were prepared with poly(A) selection (whole blood and nasal epithelium) and ribo-depletion (lung epithelium) protocols. Trans-splicing events are identified from reads spliced distantly on the same chromosome. Gene fusion events are identified from reads spliced across different chromosomes

from V(D)J recombination. After we identified all the reads with the human origin, we detected microbial reads by mapping the remaining reads onto microbial reference genomes and phylogenetic marker genes. Here, the total number of microbial reads obtained from the sample is used to estimate microbial load. We use MetaPhlAn2 (Truong et al., 2015) to assign reads on microbial marker genes and determine the taxonomic composition of the microbial communities.

Using in-house RNA-Seq data, we compare immunological and microbial profiles across asthmatics and unaffected controls for the peripheral blood, nasal, and

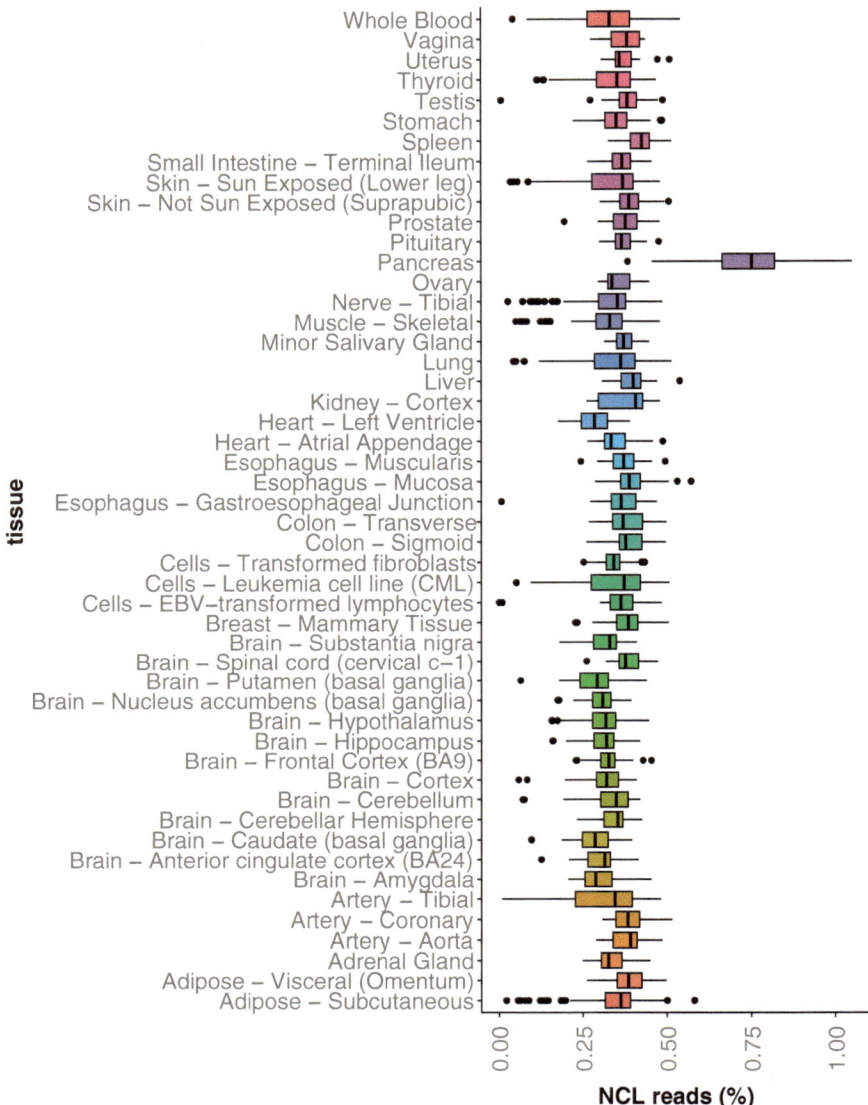

Fig. 16 Percentage of NCL reads across GTEx tissues (n = 54). Percentages are calculated from the total number of reads. Reads arising from trans-splicing, gene fusion and circRNA events are captured by a TopHat-Fusion and CIRCexplorer2 tools and reported as NCL reads

Table 5 Relative genomic abundance of microbial taxa at different levels of taxonomic classification after removal of reads with human origin (average over all samples of tissues)

Tissue	Whole blood	Nasal epithelium	Lung epithelium
N	19	19	49
Library preparation method	ploy(A) enrichment	ploy(A) enrichment	ploy(A) enrichment
Phylum			
Proteobacteria	0.0%	0.9%	100.0%
Actinobacteria	0.0%	99.1%	0.0%
Class			
Betaproteobacteria	0.0%	0.5%	86.7%
Gammaproteobacteria	0.0%	0.5%	13.3%
Actinobacteria	0.0%	98.9%	0.0%
Order			
Burkholderiales	0.0%	0.0%	87.0%
Enterobacteriales	0.0%	0.0%	12.0%
Actinomycetales	0.0%	99.5%	0.0%
Pseudomonadales	0.0%	0.5%	1.0%

Taxonomic classification is performed using MetaPhlAn2, which is able to assign the filtered unmapped reads to the microbial marker genes

large airway epithelium tissues. A total of 339 bacterial taxa were assigned with Metaphlan2 (Truong et al., 2015) across all studies.

Using Metaphlan2, we detected bacterial reads in all GTEx tissues except testis, adrenal gland, heart, brain, and nerve. We also observe no bacteria reads in the following cell lines: EBV-transformed lymphocytes (LCLs), Cells-Leukemia (CML), and Cells-Transformed fibroblasts cell lines. On average, we observe $1.43 + -0.43$ phyla assigned per sample. All samples were dominated by Proteobacteria (relative genomic abundance of $73\% + -28\%$). Other phyla detected included Acidobacteria, Actinobacteria, Bacteroidetes, Cyanobacteria, Fusobacteria, and Firmicutes. Consistent with previous studies, we observe the nasal epithelium is dominated by Actinobacteria phyla (particularly the *Propionibacterium* genus) (Yan et al., 2013), and the large airway epithelium is dominated by Proteobacteria phyla (Beck, Young, & Huffnagle, 2012) (Table 5). As a positive control for virus detection, we used GTEx samples from EBV-transformed lymphoblastoid cell lines (LCLs). ROP detected EBV virus across all LCLs samples. An example of a coverage profile of EBV virus for one of the LCLs samples is presented in Fig. 17.

We assess combinatorial diversity of the B and T cell receptor repertoires by examining the recombination of the of Variable (V) and Joining (J) gene segments from the variable region of BCR and TCR loci. We used per sample alpha diversity (Shannon entropy) to incorporate the total number of VJ combinations and their relative proportions into a single diversity metric. In order to calculate the alpha

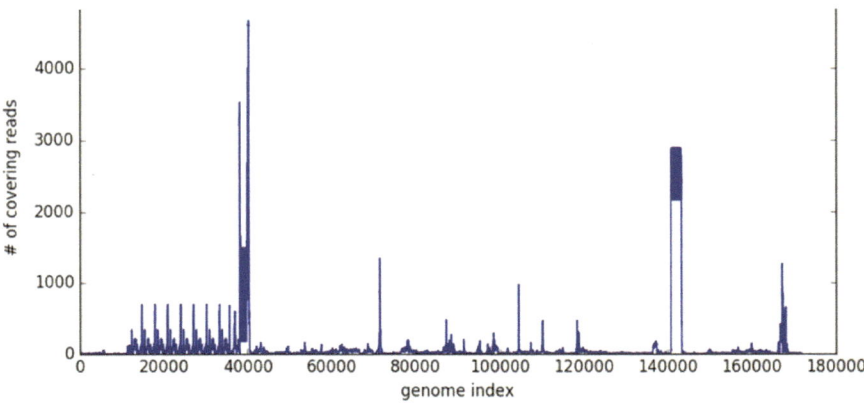

Fig. 17 An example of coverage plots of EBV virus. Viral reads were obtained by ROP protocol from GTEx RNA-Seq sample of EBV-transformed lymphoblastoid cell lines (LCLs)

diversity, we sub-sampled unmapped reads by only including reads corresponding to a sample with the smallest number of unmapped reads. Diversity within a sample was assessed using the richness and alpha diversity indices. Richness was defined as a total number of distinct events in a sample. We used Shannon Index (SI), incorporating richness and evenness components, to compute alpha diversity, which is calculated as follows:

$$SI = -\sum \left(p \times \log_2(p) \right)$$

We observed a mean alpha diversity of 0.7 among all the samples for immunoglobulin kappa chain (IGK). Spleen, minor salivary gland, and small intestine (terminal ileum) were the most immune diverse tissue, with corresponding IGK alpha diversity of 86.9, 52.05, and 43.96, respectively (Figs. 18 and 19). Across all the tissues and samples, we obtained a total of 312 VJ recombinations for IGK chains and 194 VJ recombinations for IGL chains.

Using in-house data, we investigated the effect of different library preparation techniques over the ability to detect B and T cell receptor transcripts. We compared the alpha diversity in large airway samples to nasal samples (Fig. 20). Decreased alpha diversity in large airway samples compared to nasal (2.5 for nasal versus 1.0 for large airway) could correspond to an overall decrease in percentage of immune reads. This effect can be attributed to the ribo-depletion protocol not enriching for polyadenylated antibody transcripts. Alternatively, it may result from clonal expansion of certain clonotypes responding to the cognate antigen.

Joint analysis of unmapped reads offered by ROP protocol presents several advantages over previous methods designed to examine features of unmapped reads. Our method interrogates relationships between features. To explore interactions between the immune system and microbiome, we compared immune diversity against microbial load. Microbes trigger immune responses, eliciting proliferation of antigen-

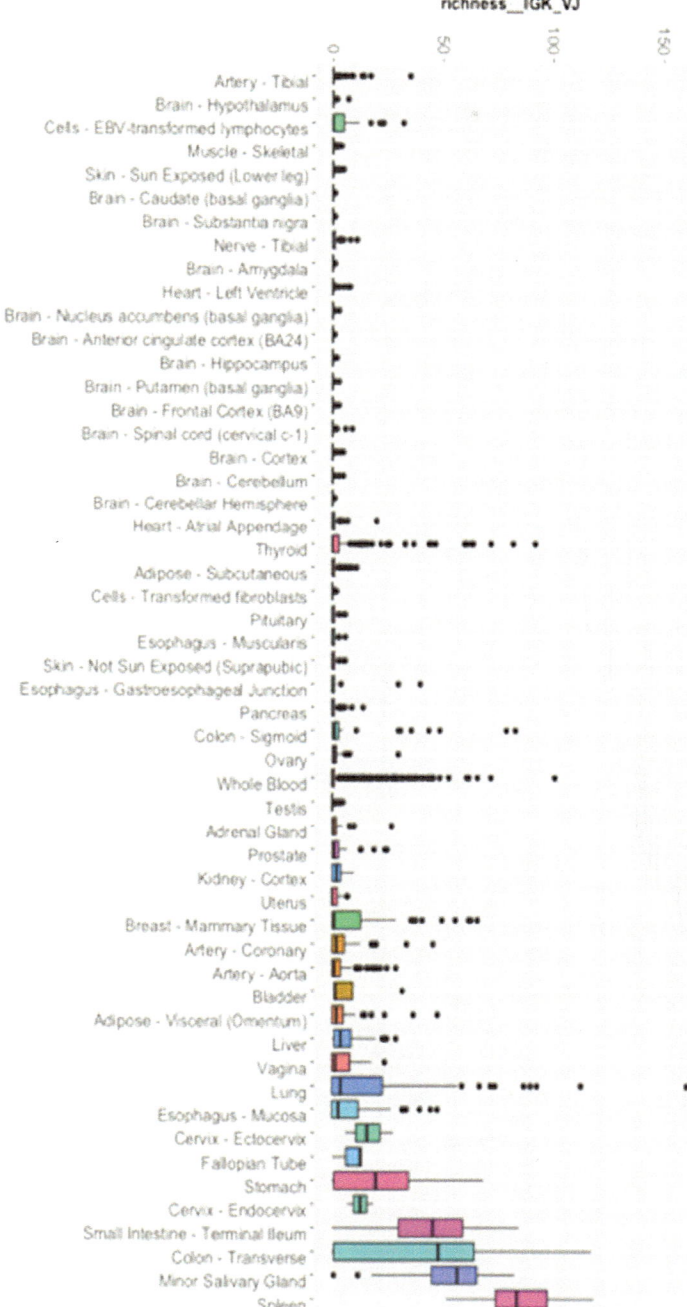

Fig. 18 Number of VJ recombinations across GTEx human tissues for IGK chain

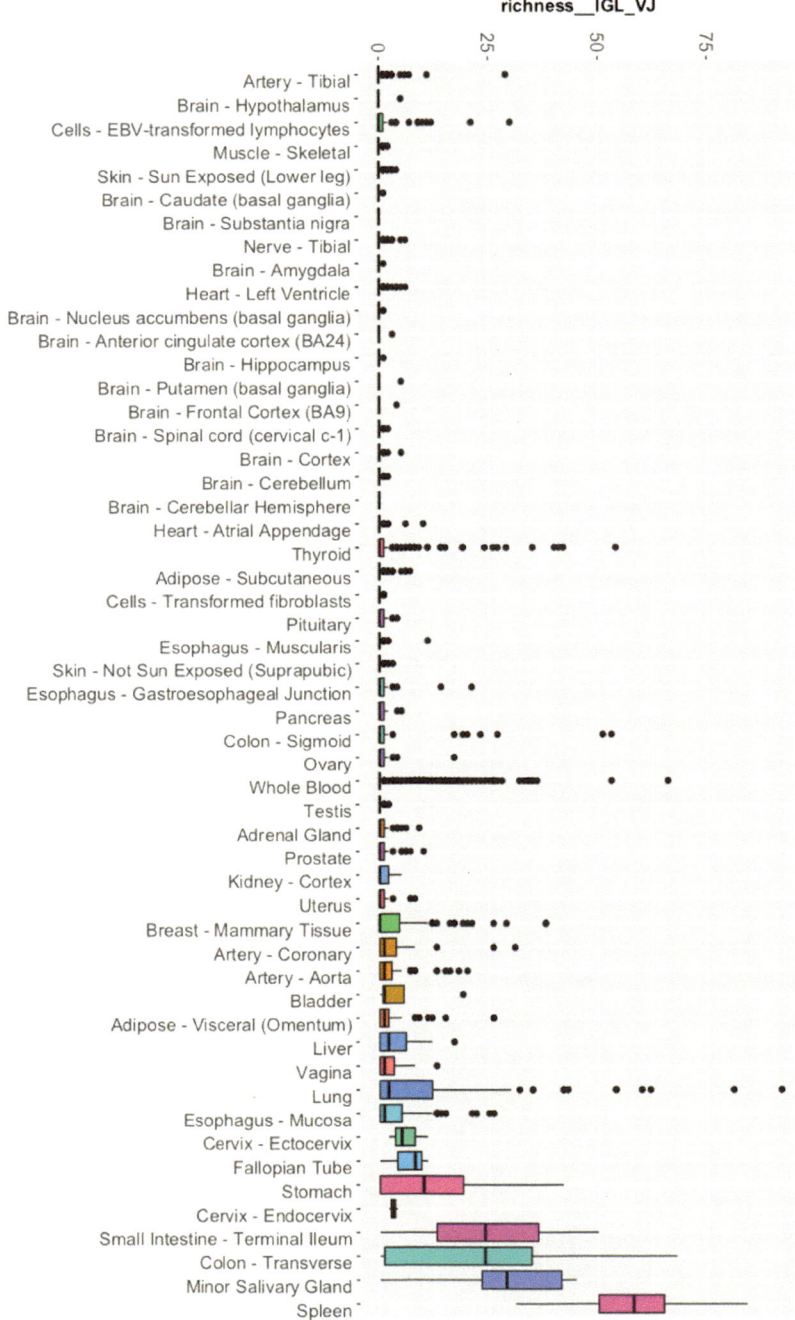

Fig. 19 Number of VJ recombinations across GTEx human tissues for IGL chain

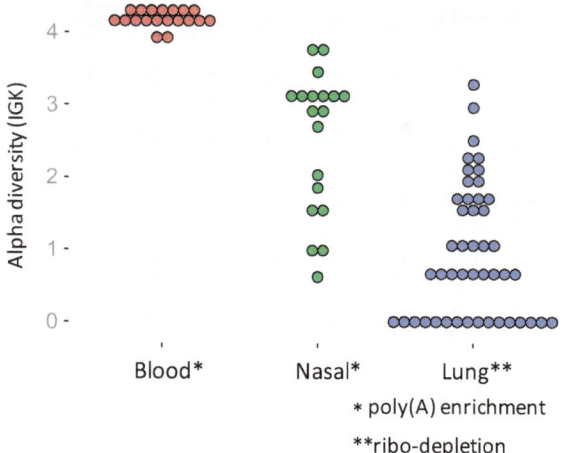

Fig. 20 Combinatorial diversity of immunoglobulin kappa locus (IGK) locus across in-house tissues. Samples were prepared by poly(A) selection (whole blood and nasal epithelium) and ribo-depletion (lung epithelium) protocols. The combinatorial diversity of IGK locus is determined based on the recombinations of the VJ gene segments. Shannon entropy measures the alpha diversity by incorporating the total number of VJ combinations and their relative proportions. Mean alpha diversity for blood samples was 4.2, for nasal samples, was 2.5, and for lung, was 1.0

specific lymphocytes. This dramatic expansion skews the antigen receptor repertoire in favor of a few dominant clonotypes and decreases immune diversity (Spreafico et al., 2016). Therefore, we reasoned that antigen receptor diversity in the presence of microbial insults should shrink. In line with our expectation, we observed that combinatorial immune diversity of IGK locus was negatively correlated with the viral load (Pearson coefficient $r = -0.55$, p-value $= 2.4 \times 10^{-6}$), consistent also for bacteria and eukaryotic pathogens across BCR and TCR loci (Fig. 21a, b, c).

Using in-house data, we compared alpha diversity of asthmatic individuals ($n = 9$) and healthy controls ($n = 10$). The combinatorial profiles of B and T cell receptors in blood and large airway tissue provide no differentiation between case control statuses. Among nasal samples, we observed decreased alpha diversity for asthmatic individuals relative to healthy controls (p-value $= 10^{-3}$) (Fig. 22b). Additionally, we used beta diversity (Sørensen–Dice index) to measure compositional similarities between samples, including gain or loss of VJ combinations of IGK locus. We calculated the beta diversity for each combination of the samples, and we produced a matrix of all pairwise sample dissimilarities. The Sørensen–Dice beta diversity index is measured as $1 - \frac{2J}{A+B}$, where J is the number of shared events, while A and B are the total number of events for each sample, respectively.

We observed higher beta diversity corresponding to a lower level of similarity across the nasal samples of asthmatic individuals in comparison to samples from unaffected controls (Fig. 22c, p-value $< 3.7 \times 10^{-13}$). Moreover, nasal samples of unaffected controls are significantly more similar than samples from the asthmatic

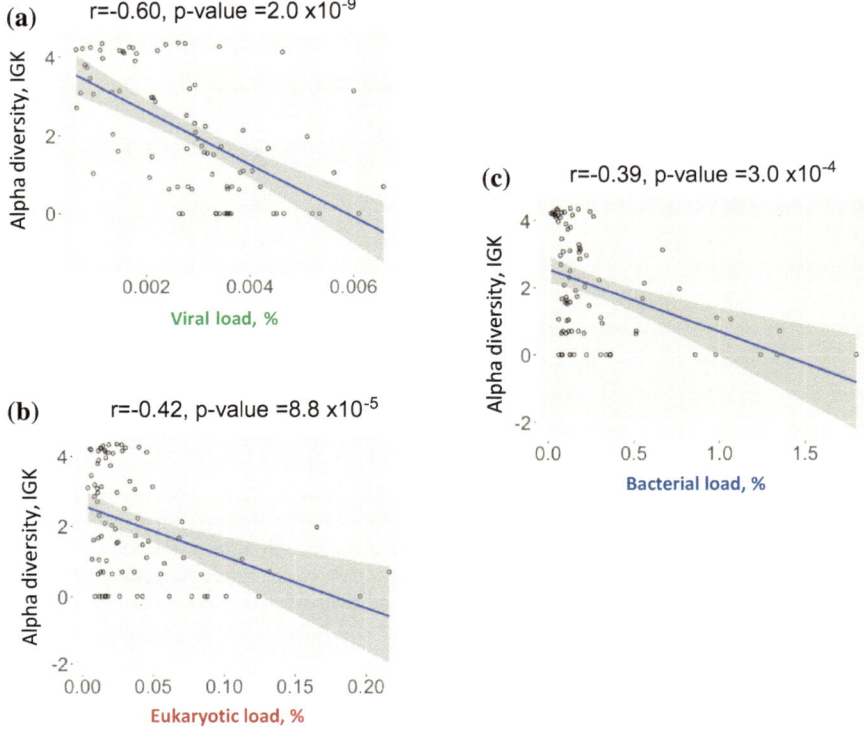

Fig. 21 Association between microbial load and immune diversity. a Scatterplot of the viral load and combinatorial immune diversity of IGK locus. Pearson correlation coefficient (r) and *p*-value are reported. **b** Scatterplot of the eukaryotic load and combinatorial immune diversity of IGK locus. Pearson correlation coefficient (r) and *p*-value are reported. **c** Scatterplot of the bacterial load and combinatorial immune diversity of IGK locus. Pearson correlation coefficient (r) and *p*-value are reported

individuals (Fig. 22c, *p*-value $< 2.5 \times 10^{-9}$). Recombination profiles of immunoglobulin lambda locus (IGL) and T cell receptor beta and gamma (TCRB and TCRG) loci yielded a similar pattern of decreased beta diversity across nasal samples of asthmatic individuals (Figs. 23, 24 and 25). Together the results demonstrate the ability of ROP to interrogate additional features of the immune system without the expense of additional TCR/BCR sequencing. Mining the unmapped reads reveals important immune and microbial information that can be leveraged for elucidating the hidden insight into human immune system. Collectively, Read Origin Protocol can identify the origin of 99.9% of RNA-Sequencing reads and excavate important underlying biological information that have been hidden in the unmapped portion of the RNA-Seq reads.

We have demonstrated the importance of hidden microbial reads buried in unmapped reads from RNA-Sequencing samples with different disease status in asthma. However, collection for such samples in disease-relevant tissue for

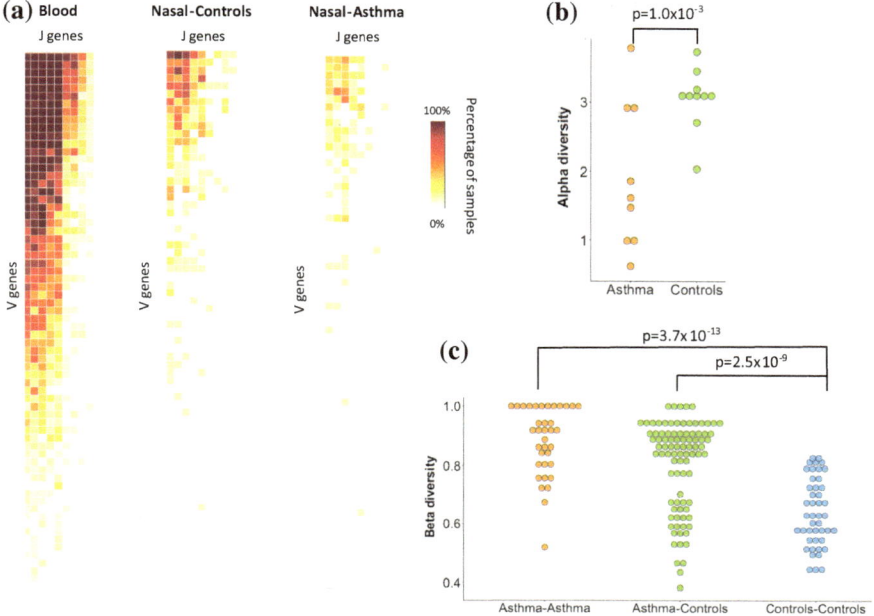

Fig. 22 Combinatorial diversity of immunoglobulin kappa locus (IGK) locus differentiates disease status. a Heatmap depicting the percentage of RNA-Seq samples supporting of particular VJ combination for whole blood (n = 19), nasal epithelium of healthy controls (n = 10), and asthmatic individuals (n = 9). Each row corresponds to a V gene, and each column correspond to a J gene. **b** Alpha diversity of nasal samples is measured using the Shannon entropy and incorporates total number of VJ combinations and their relative proportions. Nasal epithelium of asthmatic individuals exhibits decreased combinatorial diversity of IGK locus compared to healthy controls (*p*-value $= 1 \times 10^{-3}$). **c** Compositional similarities between the nasal samples in terms of gain or loss of VJ combinations of IGK locus are measured across paired samples from the same group (Asthma, Controls) and paired samples from different groups (Asthma versus Controls) using Sørensen–Dice index. Lower level of similarity is observed between nasal samples of asthmatic individuals compared to unaffected controls (*p*-value $< 7.3 \times 10^{-13}$). Nasal samples of unaffected controls are more similar to each other than to the asthmatic individuals (*p*-value $< 2.5 \times 10^{-9}$)

transcriptomic sequencing can be clinically challenging. Thus, we hypothesize that mining for the hidden treasure in blood RNA-Sequencing samples can illuminate important biological insights related to complex diseases. Other than in cases of sepsis, we currently lack a comprehensive understanding of the human microbiome in blood, as blood has been generally considered a sterile environment lacking proliferating microbes (Drennan, 1942). However, over the past few decades, this assumption has been challenged (McLaughlin et al. 2002; Nikkari, McLaughlin, Bi, Dodge, & Relman, 2001) and the presence of a microbiome in the blood has received increasing attention (Amar et al., 2011; Sato et al., 2014; Païssé et al., 2016).

To explore potential connections between the microbiome and diseases of the brain, we performed a comprehensive analysis of microbial products detected in

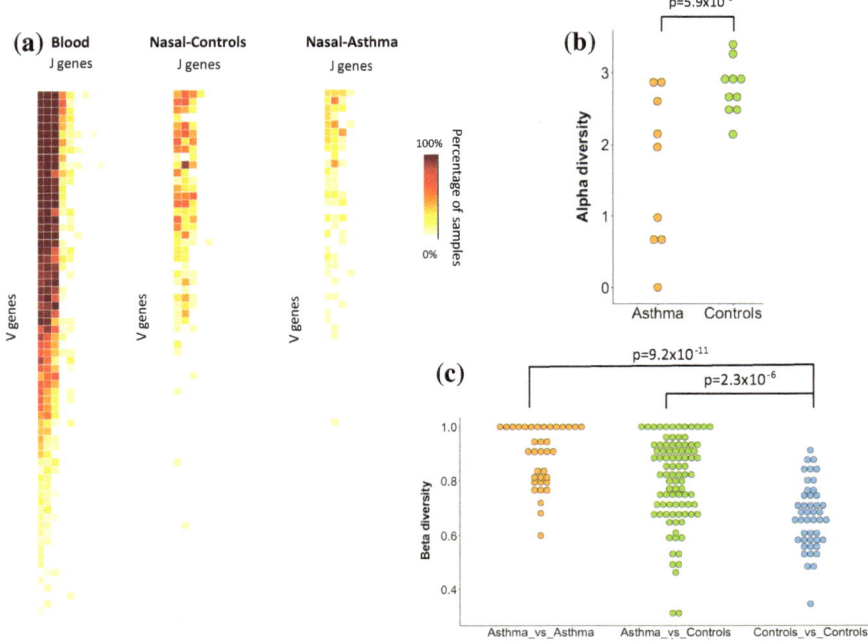

Fig. 23 Combinatorial diversity of immunoglobulin lambda locus (IGL) locus differentiates disease status. a Heat map depicting the percentage of RNA-Seq samples supporting particular VJ combination for whole blood, nasal epithelium of healthy controls and asthmatic individuals. Each row corresponds to a V gene and each column corresponds to a J gene. **b** Alpha diversity is measured using the Shannon entropy incorporating the total number of VJ combinations and their relative proportions. Nasal epithelium of asthmatic individuals exhibits decreased combinatorial diversity of IGK locus compared to that of healthy controls (p-value $= 5.9$ x 10^{-3}). **c** Compositional similarities between the samples in terms of gain or loss of VJ combinations of IGK locus are measured using the Sørensen–Dice index across pairs of samples from the same group (Asthma, Controls) and pairs of samples from different groups (Asthma versus Controls). Lower level of similarity is observed between nasal samples of the asthmatic individuals compared to the unaffected controls (p-value $<$ 9.2×10^{-11}). Nasal samples of the unaffected controls are more similar to each other than to the asthmatic individuals (p-value $< 2.3 \times 10^{-6}$)

blood in almost two hundred individuals, including patients with schizophrenia, bipolar disorder and sporadic amyotrophic lateral sclerosis. These three disease groups represent complex polygenic traits that affect the central nervous system with largely unknown etiology. Moreover, roles for the microbiome in all the diseases have been previously hypothesized (Foster & Neufeld, 2013; Wu et al., 2015; Dinan, Borre, & Cryan, 2014; Dickerson, Severance, & Yolken, 2017). We used available high-quality RNA sequencing (RNA-Seq) reads from whole blood that fail to map to the human genome as candidate microbial reads for microbial classification. We observed an increased diversity of microbial communities in schizophrenia patients, and we replicated this finding in an independent dataset. Careful analyses, including the use of

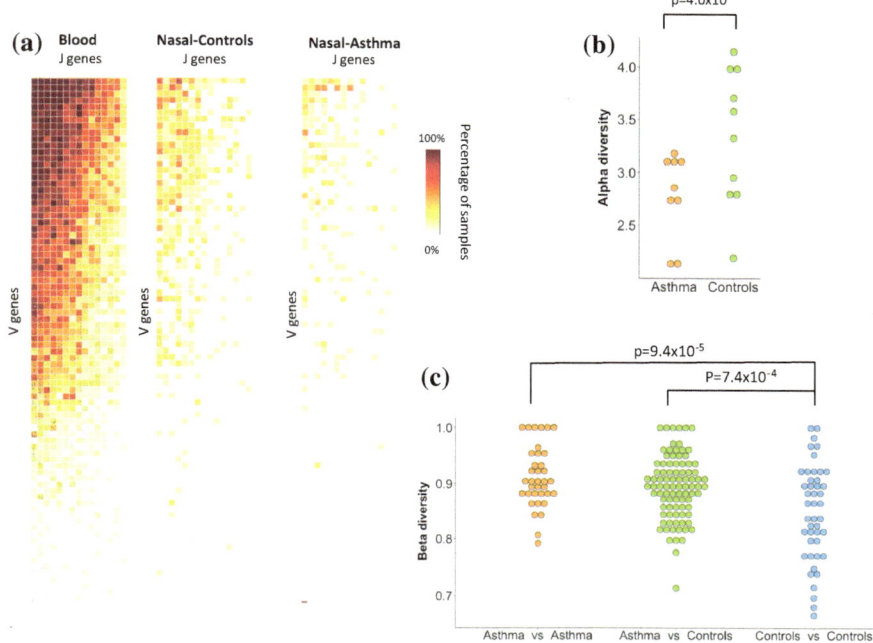

Fig. 24 Combinatorial diversity of T cell receptor beta (TCRB) locus differentiates disease status. a Heat map depicting the percentage of RNA-Seq samples supporting of particular VJ combination for whole blood, nasal epithelium of healthy controls and of asthmatic individuals. Each row corresponds to a V gene and each column corresponds to a J gene. **b** Alpha diversity is measured using the Shannon entropy incorporating the total number of VJ combinations and their relative proportions. The nasal epithelium of asthmatic individuals exhibits a decrease in combinatorial diversity of IGK locus compared to that of healthy controls (p-value $= 4.0 \times 10^{-2}$). **c** Compositional similarities between the samples in terms of gain or loss of VJ combinations of IGK locus are measured using the Sørensen–Dice index across pairs of samples from the same group (Asthma, Controls) and pairs of sample from different groups (Asthma versus Controls). Lower level of similarity is observed between nasal samples of asthmatic individuals compared to unaffected controls (p-value $< 9.4 \times 10^{-5}$). Nasal samples of unaffected controls are more similar to each other than to the asthmatic individuals (p-value $< 7.4 \times 10^{-4}$)

positive and negative control datasets, suggest that these detected phyla represent true microbial communities in whole blood and are not present in samples due to contaminants. With the increasing number of RNA-Seq data sets, our approach may have great potential for application across different tissues and disease types.

Firstly, blood samples were obtained for RNA-Sequencing and divided into discovery and replication sample groups. The discovery sample consists of unaffected controls (Controls, n = 49) and patients with three brain-related disorders: schizophrenia (SCZ, n = 48), amyotrophic lateral sclerosis (ALS, n = 47) and bipolar disorder (BPD, n = 48). The replication sample includes Controls (n = 88) and SCZ samples (n = 91). For the discovery sample, RNA-Seq libraries were prepared

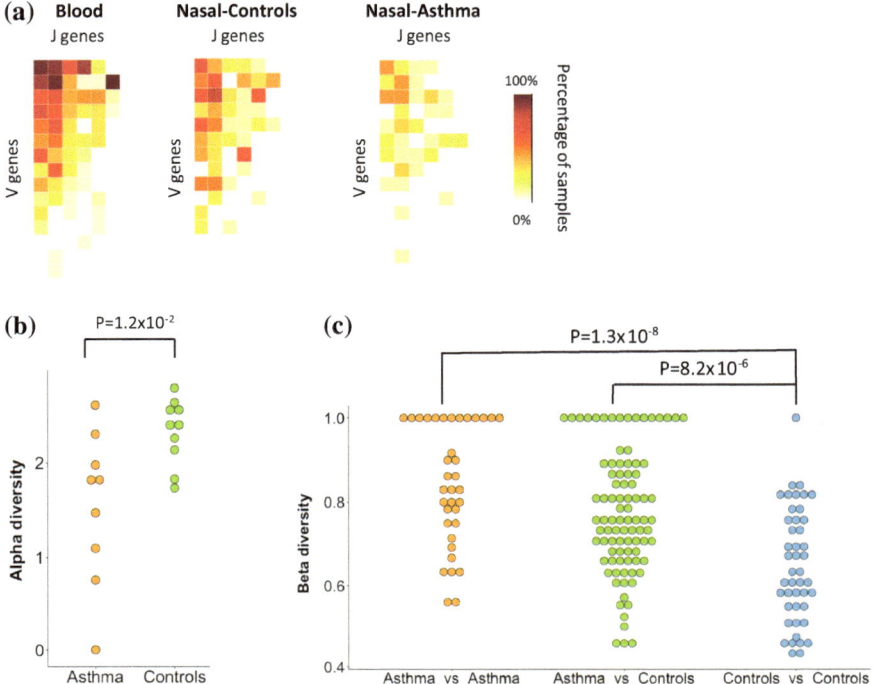

Fig. 25 **Combinatorial diversity of T cell receptor gamma (TCRG) locus differentiates disease status**. **a** Heat map depicting the percentage of RNA-Seq samples supporting of a particular VJ combination for whole blood, nasal epithelium of healthy controls and asthmatic individuals. Each row corresponds to a V gene and each column corresponds to a J gene. **b** Alpha diversity is measured using the Shannon entropy incorporating the total number of VJ combinations and their relative proportions. Nasal epithelium of asthmatic individuals exhibits decreased combinatorial diversity of IGK locus compared to that of healthy controls (p-value $= 1.2 \times 10^{-2}$, ANOVA). **c** Compositional similarities between the samples in terms of gain or loss of VJ combinations of IGK locus are measured using the Sørensen–Dice index across pairs of samples from the same group (Asthma, Controls) and pairs of samples from different groups (Asthma versus Controls). Lower level of similarity is observed between nasal samples of asthmatic individuals compared to unaffected controls (p-value $< 1.3 \times 10^{-8}$,). Nasal samples of unaffected controls are more similar to each other than to the asthmatic individuals (p-value $< 8.2 \times 10^{-6}$)

using Illumina's TruSeq RNA v2 protocol, including ribo-depletion protocol (Ribo-Zero Gold). In total, we obtained 6.8 billion 2×100 bp paired-end reads for the primary study ($35.3M \pm 6.0$ paired-end reads per sample). The replication sample was processed at the same core facility using the same standardized procedures as the discovery sample. However, the RNA-Seq libraries were prepared with poly(A) enrichment, a procedure more selective than the total RNA that was used for the discovery sample. A total of 3.8 billion reads were obtained ($26.3M \pm 12.0$ paired-end reads per sample).

We separated human and non-human reads, and use the latter as candidate microbial reads for taxonomic profiling of microbial communities. To identify potentially microbial reads, we developed an analysis pipeline tailored from Read Origin Protocol. First, we filtered read pairs and singleton reads mapped to the human genome or transcriptome, as noted in ROP process. However, unlike ROP, we performed normalization by sub-sampling to 100,000 reads for each sample, because total number of reads may affect microbial profiling. Next, we filtered out low-quality and low-complexity reads using FASTX (http://hannonlab.cshl. edu/fastx_toolkit/) and SEQCLEAN ("https://sourceforge.net/projects/seqclean/," n.d.). Finally, the remaining reads were realigned to the human references using the Megablast aligner (Camacho et al., 2009) in order to exclude any potentially human reads, same as Read Origin Protocol procedures. The remaining reads were used as candidate microbial reads in detailed subsequent analyses.

To access the assembly and richness of the microbial RNA in blood, we used phylogenetic marker genes to assign the candidate microbial reads to the bacterial and archaeal taxa. We used PhyloSift to perform taxonomic profiling of the whole blood samples (Darling et al., 2014). PhyloSift makes use of a set of protein coding genes found to be relatively universal (i.e., present in nearly all bacterial and archaeal taxa) and have low variation in copy number between taxa. Homologs of these genes in new sequence data (e.g., the transcriptomes used here) are identified and then placed into a phylogenetic and taxonomic context by comparison to references from sequenced genomes. For our replication study, we used MetaPhlAn for microbial profiling v.1.7.7 (Segata et al., 2012). MetaPhlAn was run in 2 stages; the first stage identifies the candidate microbial reads (i.e., reads hitting a marker) and the second stage profiles meta-genomes in terms of relative abundances. We used MetaPhlAn, rather than PhyloSift, due to differences in library preparation (polyA enrichment versus Ribo-Zero); there were an insufficient number of reads matching the database of the marker genes curated by PhyloSift for adequate microbial profiling of the replication sample.

To study the composition of microbial RNA in blood, we determined the microbial meta-transcriptome present in the blood of unaffected controls (Controls, n = 49) and patients with three brain-related disorders: schizophrenia (SCZ, n = 48), amyotrophic lateral sclerosis (ALS, n = 47) and bipolar disorder (BPD, n = 48) (Fig. 26; Table 1). Using our ROP pipeline, an average of 33,546 of 100,000 unmapped reads are identified as high quality, unique non-host reads and were used as candidate microbial reads in our analyses. From these, PhyloSift was able to assign an average of 1235 reads (1.24% ± 0.41%, mean ± standard deviation) to the bacterial and archaeal gene families. A total of 1880 taxa were assigned, with 23 taxa at the phylum level (Fig. 27). Most of the taxa we observed derived from bacteria (relative genomic abundance 89.8 ± 7.4%), and a smaller portion derived from archaea (relative genomic abundance 12.28 ± 6.4%).

In total, we observed 23 distinct microbial phyla with on average 4.1 ± 2.0 phyla per individual. The large majority of taxa observed in our sample is not universally present in all individuals; the single exception is Proteobacteria, which dominates all samples with 73.4 ± 18.3% relative abundance (Fig. 27, dark green color). Several

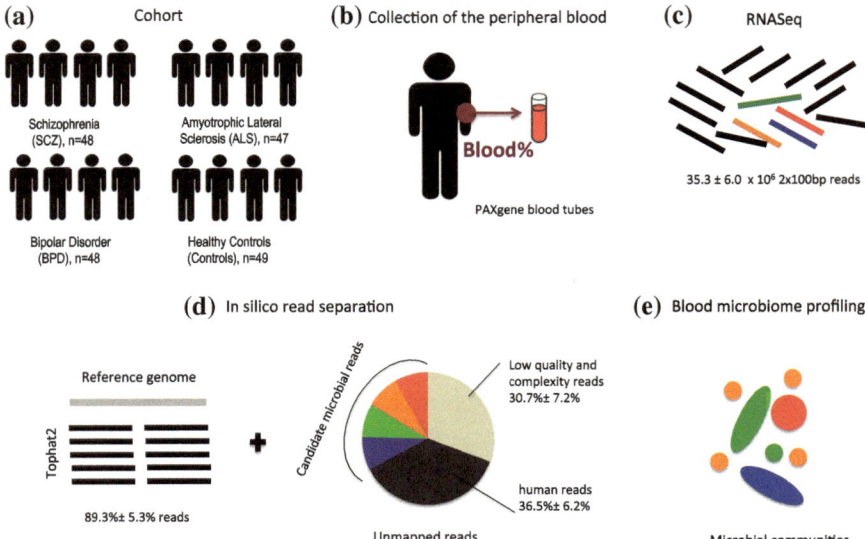

Fig. 26 Microbial profiling using RNA-Seq data from whole blood. **a** We analyzed a cohort of 192 individuals from four subject groups, i.e. Schizophrenia (SCZ, n = 48), amyotrophic lateral sclerosis (ALS n = 47), bipolar disorder (BPD n = 48), unaffected control subjects (Controls n = 49). **b** Peripheral blood was collected for RNA collection. **c** RNA-Seq libraries were prepared from total RNA using ribo-depletion protocol. **d** Reads that failed to map to the human reference genome and transcriptome were sub-sampled and further filtered to exclude low-quality, low complexity, and remaining potentially human reads. **e** High quality, unique, non-host reads were used to determine the taxonomic composition and diversity of the detected microbiome

bacterial phyla show a broad prevalence across individuals and disorders (present in 1/4 of the samples of each subject group). Those phyla include Proteobacteria, Firmicutes and Cyanobacteria, with relative abundance 73.4 ± 18.3%, 14.9 ± 10.9%, and 11.0 ± 8.9%. This is in line with recent published work on the blood microbiome using 16S targeted metagenomic sequencing reporting relative abundance of 80.4–87.4% and 3.0–6.4% for Proteobacteria and Firmicutes, respectively (Païssé et al., 2016). The other two phyla identified in this study (Actinobacteria and Bacteroidetes) were also detected in our sample in more than 25 individuals. Although Proteobacteria and Firmicutes are commonly associated with the human microbiome (Human Microbiome Project, 2012), some members of these phyla might be associated with reagent and environmental contaminants (Salter et al., 2014; Strong et al., 2014). Thus, it is necessary to address the potential contaminations from experimental protocols prior to assessing the clinical prevalence of microbiome in blood.

To validate our pipeline and investigate the possibility of contamination introduced during RNA isolation, library preparation, and sequencing steps, we performed both negative and positive control experiments. We performed the following negative control experiment to investigate the possibility of DNA contamination introduced during RNA isolation, library preparation and sequencing steps. We applied our

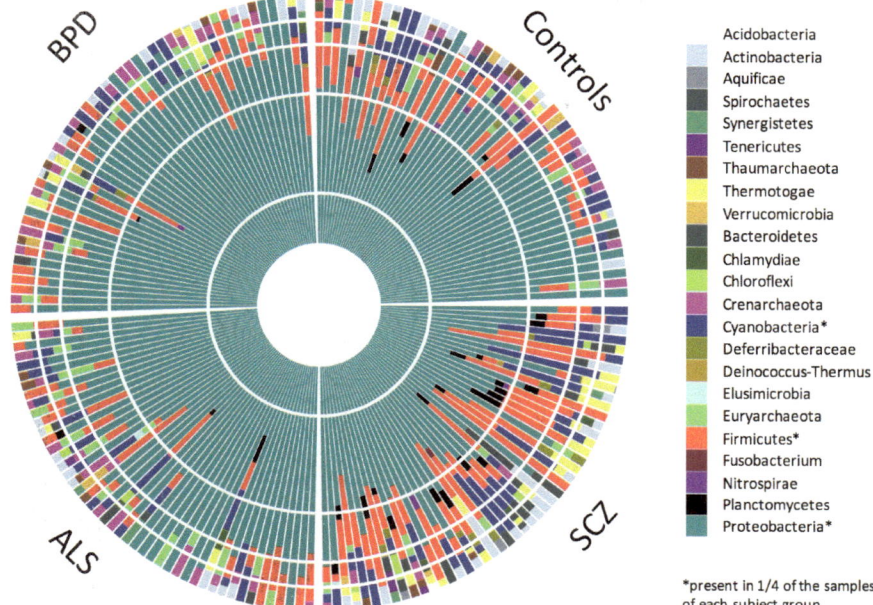

Fig. 27 Relative abundances of microbial taxa at phylum level. Phylogenetic classification is performed using PhyloSift, which is able to assign the filtered candidate microbial reads to the microbial genes from 23 distinct taxa on the phylum level

pipeline to RNA-Seq reads from six B-lymphoblast cell line (LCLs) samples. LCLs are cultured in a sterile tissue culture environment using aseptic techniques. We therefore expect these to lack any presence of microbial species. Neither PhyloSift nor MetaPhlAn detected bacterial or archaeal microorganisms in the LCLs samples (see Table 6c). This experiment also serves as a positive control, as the only virus detected by PhyloSift is the Epstein-Barr virus, used for transfection and transformation of lymphocytes to lymphoblasts (Abu-Shanab & Quigley, 2010).

We used a more direct positive control dataset to validate the feasibility of using human RNA-Seq for detection of microbial organisms and applied the Read Origin Protocol pipeline to RNA-Seq data collected from epithelial cells infected with Chlamydiae (Humphrys et al. 2013). The authors collected data using ribo-depletion and polyA selection protocols at 1 and 24 h post infection. PhyloSift was able to detect the Chlamydiae phylum in 100,000 reads randomly subsampled from unmapped reads, confirming the validity of the bioinformatics pipeline used (Table 7).

The design of experimental procedures such as blood draw and subsequent downstream lab procedures may lead to global contamination effects. In our data, there is minimal evidence that the detected microbial communities are confounded by contamination due to experimental procedures. First, all RNA samples were subjected to the same standardized RNA isolation protocols, library preparation and sequencing procedures. With the exception of Proteobacteria, which has been reported to be the

Table 6 Data overview

(a) Primary study. Whole Blood RNA-3Seq study of 192 samples across four subject
groups (controls, ALS, BPD, SCZ)

Disease status	Control	SCZ	BPD	ALS	Total
N	49	48	48	47	192
Number of read pairs, mean (std), millions	36.7 (6.1)	30.2 (6.1)	37.9 (4.8)	36.2 (3.4)	35.3 (6.0)
Number of mapped read pairs, mean (std), millions	26.7 (5.8)	19.7 (6.3)	29.2 (4.1)	27.7 (3.4)	25.9 (6.2)
Number of singletons[a] reads, mean (std), millions	6.2 (1.5)	6.2 (1.4)	5.5 (0.9)	5.3 (1)	5.8 (1.3)
Number of unmapped pairs, mean (std), millions	3.8 (2.0)	4.3 (1.2)	3.3 (1)	3.1 (1.6)	3.6 (1.6)

(b) Positive control. Whole blood exome sequencing from two samples

SampleID	088503	088758A
Number of single-end reads	62,813,827	62,912,604
Number of unmapped reads	1,190,675	1,189,314

(c) Negative control. RNA-Seq of B-lymphoblast cell line. Samples were collected from a
trio (father, mother, offspring) in duplicate, total number of sequenced samples is 6

Individual ID	SampleID	Number of read pairs	Number of mapped read pairs	Number of singletons[a]	Number of unmapped pairs[b]
GM12740A	13R1700	23,629,424	20,989,748	909,710	1,729,966
GM12740A	13R1697	23,045,351	20,863,402	808,997	1,372,952
GM12750B	13R1702	23,823,173	21,761,316	750,061	1,311,796
GM12750B	13R1699	25,341,063	15,377,305	4,121,435	5,842,323
GM12751A	13R1701	23,710,989	21,334,518	879,394	1,497,077
GM12751A	13R1698	25,054,808	22,543,667	920,777	1,590,364

(d) Replication study. Whole Blood RNA-Seq study of 192 samples across two subject
groups (Controls, SCZ)

Disease status	Control	SCZ	Total
N	88	91	179
Number of read pairs, mean (std), millions	30.1 (13.2)	27.6 (11.8)	26.3 (12.0)
Number of mapped read pairs, mean (std), millions	24.1 (8.6)	19 (7.9)	20.8 (8.2)
Number of singletons[a] reads, mean (std), millions	2.6 (1.2)	2.4 (1.5)	2.3 (1.3)
Number of unmapped pairs, mean (std), millions	2.6 (5.0)	4.9 (7.7)	3.2 (6.2)

[a]Reads with one end mapped another end unmapped
[b]Both ends of the paired-end read are unmapped

Table 7 Relative abundance of the Chlamydiae phylum at 1 and 24 h post infection

Time	polyA RNA-Seq (%)	Ribo-Zero RNA-Seq (%)
1 h	0	0.1
24 h	24	0.9

Genomic abundances of Chlamydiae phylum for RNA-Seq samples prepared by ribo-depletion and polyA selection protocols at 1 and 24 h post infection. Phylogenetic classification is performed by PhyloSift for 100,000 reads randomly subsampled from unmapped reads

most abundant phylum in whole blood (Cho & Blaser, 2012), we observe no phylum present in all individuals, suggesting the absence of a uniform contaminator due to experimental procedures applied across all samples.

Second, we collected two blood tubes per individual, of which one was randomly chosen for subsequent RNA sequencing. If skin contamination upon first blood draw occurs, due to contact with the needle, its effect will be randomly distributed across half of individuals in our cohort and should therefore not affect downstream between-group analyses.

Third, it is vital to scrutinize the potential impact of parameters that are variable between samples, such as experimenter (i.e., lab technician who extracted RNA from blood collections) (Greenblum et al., 2012). To investigate these potential effects, we grouped samples by various experimental variables, including sequencing run and experimenter. We observed no evidence that the detected microbial communities are confounded by contamination, which agrees with previously reported studies that documented at most a low background signal introduced by such variables (Cho & Blaser, 2012) (see also Figs. 28 and 29). In addition, we included all available technical covariates, such as RNA integrity number (RIN), batch, flow cell lane and RNA concentration, in our disease specific analyses.

Finally, an independent technology was used to validate the detected microbial composition in our RNA-Seq cohort. We used available blood whole exome sequence data from two individuals from the cohort (see Table 6b). We applied the Read Origin Protocol pipeline and compared results from both technologies. Despite the use of different technologies and reagents, microbiome profiles from both sequencing procedures were found to be in close agreements. For both individuals, we were able to detect several microbial phyla, all of which were also identified using RNA-Seq. Conversely, RNA-Seq was able to detect several microbial phyla not detected using exome sequencing (333) (Table 8). Taken together, these results confirm the validity and potential of our tailored Read Origin Protocol pipeline.

To summarize, no microbiome sequences were detected in transcriptome data in lymphoblast cell lines (negative control), and we only detected the Chlamydiae phylum in RNA-Seq from cells infected with Chlamydiae (positive control). We examined experimental procedures and technical parameters on microbial composition, and we observed no link between the presence of microbial communities and possible confounders. Thus, we assume any microbial reads found in subsequent analysis is originated from patients' microbiome.

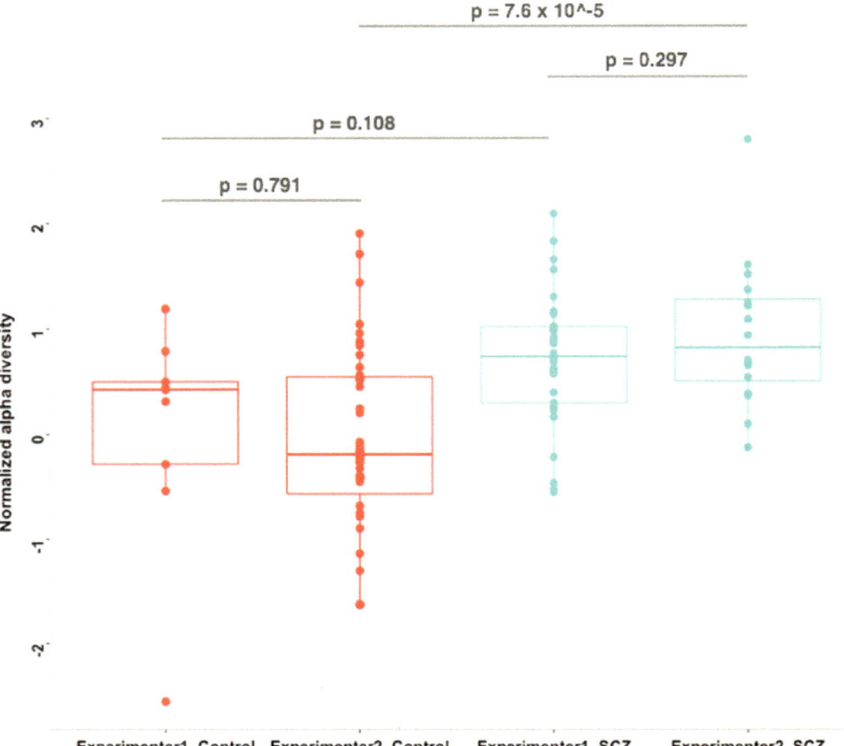

Fig. 28 Experimenter and alpha diversity measures. We analyzed whether experimenter (lab technician) can account for differences in alpha diversity between controls and schizophrenia. We find no evidence that experimenter has an effect. We included all samples for which this information was available (sample sizes: Experimenter1_Control = 9, Experimenter2_control = 39, Experimenter1_SCZ = 29, Experimenter2_SCZ = 16). Student t-test was used to evaluate differences between groups, and p-values are reported

To compare the inferred microbial composition found in blood with that in other body sites, we used taxonomic composition of 499 meta-genomic samples from Human Microbiome Project (HMP) obtained by MetaPhlAn or five major body habitats (gut, oral, airways, and skin) (Human Microbiome Project, 2012). Of the 23 phyla discovered in our samples, 15 were also found in HMP samples, of which 13 are confirmed by at least ten samples. Our data suggest that the predominant phyla detected in blood are most closely related to the known oral and gut microbiome (Table 9). Comparing the microbial composition of whole blood with the microbiome detected in atherosclerotic plaques (Koren et al., 2010), we observed that the four phyla that together make up for >97% of the microbiome in plaques are also identified in our sample (Firmicutes, Bacteroidetes, Proteobacteria, and Actinobacteria).

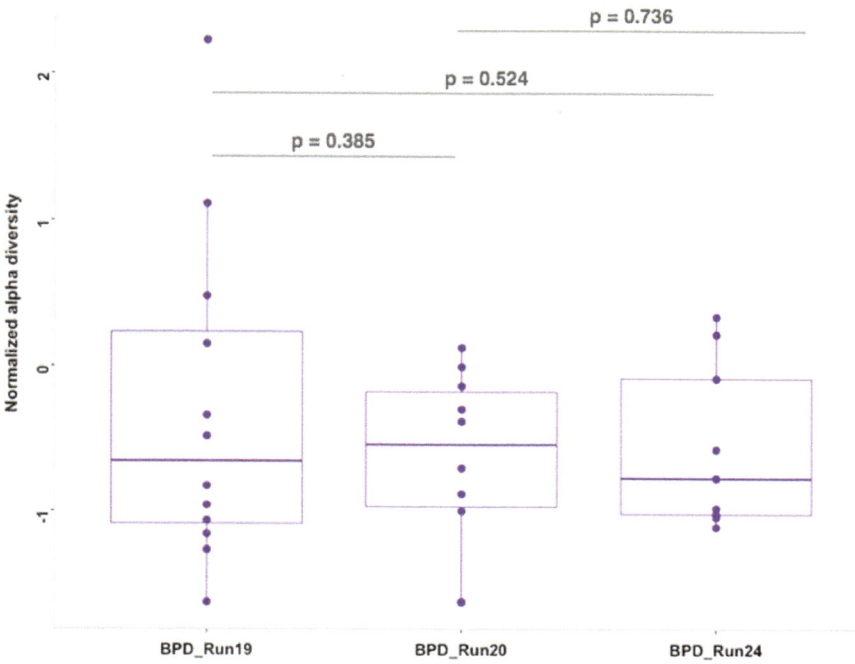

Fig. 29 RNA isolation run and alpha diversity measures. We assessed whether alpha diversity measures vary across isolation runs. We included only runs for which at least 8 samples were included (sample sizes: Run19 = 12, Run20 = 10, Run24 = 9) and found no evidence for this. Student t-test was used to evaluate differences between runs, and *p*-values are reported

	Sample	Microbial phyla	Exome sequencing	RNA-seq
Table 8 Cross Technology Validation of RNA-Seq versus Exome sequencing	Sample1	Firmicutes	✔	✔
	Sample1	Acidobacteria	✔	✔
	Sample1	Proteobacteria	✔	✔
	Sample1	Cyanobacteria		✔
	Sample1	Aquificae		✔
	Sample1	Planctomycetes		✔
	Sample2	Proteobacteria	✔	✔
	Sample2	Firmicutes	✔	✔
	Sample2	Thermotogae		✔
	Sample2	Cyanobacteria		✔
	Sample2	Chlamydiae		✔
	Sample2	Aquificae		✔
	Sample2	Verrucomicrobia		✔

Table 9 Similarities of blood microbiome profiles with other body sites

Microbial Phyla	Number of RNASeq samples	Relative abundance mean (std)	Present in more then 50% of samples in HMP	Number HMP samples taxa is detected in (total number of sample is 499)	Reference suggesting the presence of the phylum on/in human body
Fusobacterium	1	0.041 (0.000)	Oral, stool	309	Human Microbiome Project (2012)
Elusimicrobia	3	0.026 (0.012)	N/A	0	Oral, Adler et al. (2013)
Acidobacteria	3	0.032 (0.011)	N/A	3	Gut, Andersson, Anders et al. (2014)
Tenericutes	3	0.051 (0.030)	No	14	Human Microbiome Project (2012)
Deinococcus-Thermus	4	0.025 (0.012)	N/A	0	Gut, Lagier, Jean-Christophe et al. (2012)
Synergistetes	5	0.036 (0.025)	N/A	0	N/A
Aquificae	5	0.057 (0.036)	N/A	2	Gut, Lagier, Jean-Christophe, et al. (2012)
Nitrospirae	6	0.039 (0.014)	N/A	0	N/A
Spirochaetes	6	0.037 (0.020)	Oral	143	Human Microbiome Project (2012)
Chlamydiae	8	0.038 (0.023)	Oral, stool	249	Human Microbiome Project (2012)
Chloroflexi	9	0.045 (0.025)	No	35	Human Microbiome Project (2012)
Verrucomicrobia	11	0.048 (0.033)	No	65	Human Microbiome Project (2012)

(continued)

Table 9 (continued)

Microbial Phyla	Number of RNASeq samples	Relative abundance mean (std)	Present in more then 50% of samples in HMP	Number HMP samples taxa is detected in (total number of sample is 499)	Reference suggesting the presence of the phylum on/in human body
Deferribacteraceae	14	0.044 (0.014)	Oral, stool	248	Human Microbiome Project (2012)
Thaumarchaeota	18	0.041 (0.015)	N/A	0	Skin, Probst, Alexander (2013)
Bacteroidetes	25	0.067 (0.047)	Oral, stool, skin	375	Human Microbiome Project (2012)
Planctomycetes	30	0.060 (0.029)	N/A	0	Gut, Cayrou, Caroline et al. (2013)
Thermotogae	35	0.064 (0.036)	N/A	0	Gut, Kaoutari et al. (2013)
Euryarchaeota	48	0.072 (0.035)	No	77	Human Microbiome Project (2012)
Crenarchaeota	55	0.057 (0.025)	N/A	0	Gut, Turnbaugh et al. (2006)
Actinobacteria	56	0.057 (0.047)	Airways, oral, stool, skin	438	Human Microbiome Project (2012)
Cyanobacteria[a]	94	0.110 (0.089)	No	14	Human Microbiome Project (2012)
Firmicutes[a]	140	0.149 (0.109)	Airways, oral, stool, skin	450	Human Microbiome Project (2012)
Proteobacteria[a]	190	0.734 (0.183)	Airways, oral, stool, skin	399	Human Microbiome Project (2012)

[a]Phyla are present in at least 1/4th of the samples in each subject group

Our data suggest that the predominant phyla of the blood microbiome are most closely related with the known oral and gut microbiome. That is, out of eight blood microbiome phyla detected in at least 50% of HMP samples, all are found in the oral or stool samples. Among those four phyla (Spirochaete, Deferribacteres, Chlamydiae, and Fusobacterium) are found in the majority of the oral or stool samples, but in less than 50% of other tissues. The Bacteroidetes phylum was present in the majority of the oral, stool and skin samples. The remaining three phyla (Actinobacteria, Firmicutes, and Proteobacteria) were present in the majority of HMP samples across all tissues. The majority of the eight phyla not confirmed by HMP have been described to be present in or on the human body. For example, Crenarchaeota, Thermotogae, Deinococcus-Thermus and Planctomycetes have been associated with the human gut microbiome (Turnbaugh et al., 2008; El Kaoutari, Armougom, Gordon, Raoult, & Henrissat, 2013; Lagier, Million, Hugon, Armougom, & Raoult, 2012; Cayrou, Sambe, Armougom, Raoult, & Drancourt, 2013), while other phyla are reported to be present in other human tissues (e.g., Thaumarchaeota, which is present in skin microbiome). However, it should be noted that the sequencing technology does not allow for identification of the origin of microbial RNA. That is, we cannot distinguish whether the observed microbial signatures in blood are originate from bacterial communities actually present in the blood, or whether the RNA crossed into the blood stream from elsewhere. Nevertheless, the microbial reads detected in blood can reflect microbiomes in patients and can be investigated for possible relationship with complex diseases in brain.

Microbial diversity, or alpha diversity, within each sample was determined using the inverse Simpson index. This index simultaneously assesses both richness (corresponding to the number of distinct taxa) and relative abundance of the microbial communities within each sample (Simpson, 1949). In particular, it enables effective differentiation between the microbial communities shaped by the dominant taxa and the communities with many taxa with even abundances (Whittaker, 1972) (*asbio* R package). To measure sample-to-sample dissimilarities between microbial communities, we use Bray-Curtis beta diversity index, which accounts for both changes in the abundances of the shared taxa and for taxa uniquely present in one of the samples (*vegan* R package). Higher beta diversity indicates higher level of dissimilarity between microbial communities, providing a link between diversity at local scales (alpha diversity) and the diversity corresponding to total microbial richness of the subject group [gamma diversity (Koleff, Gaston, & Lennon, 2003)].

To test for differences in alpha diversity between disease groups, we fit an analysis of covariance (ANCOVA) model using normalized values of alpha, including sex and age, and technical covariates (RNA INtegrity value (RIN), batch, flow cell lane and RNA concentration) into the model. Bonferroni correction for multiple testing was used. To determine the relative effect size of alpha diversity on schizophrenia status, we fit a logistic regression model including the same covariates and measure reduction in R^2 comparing the full logistic regression model versus a reduced model with alpha removed.

To test for differences in alpha diversity between disease groups, we fit the following analysis of covariance (ancova) model

$$alpha\_norm \sim Sex + Age + Technical\ covariates + Disease\ status,$$

where $Alpha\_norm$ = alpha values after inverse normal transformation and Age = Individual's age at blood draw. Technical covariates include: RIN, $Batch$ $(Plate\_number)$, $Concentration$, and $Flow\ cell\ lane$, where RIN = RNA integrity value, a measure for RNA quality and $Concentration$ = RNA concentration prior to normalization at the genotyping core. The effect of disease status was estimated by first regressing out the effects of the included covariates. Adjustment for pairwise comparisons for all possible disease status pairs (6 comparisons) was performed using Bonferroni correction for multiple testing.

To determine the relative effect size of alpha diversity on schizophrenia status, we fit the following logistic regression model:

$$SCZ \sim Sex + Age + Technical\ covariates + alpha\_norm$$

where SCZ is a binary variable, which is coded as true if the sample belongs to the SCZ cohort.

Variation explained is by $alpha\_norm$ was measured by the reduction in R^2, comparing the full logistic regression model versus a reduced model with $alpha\_norm$ removed. We separately tested for differences in alpha by sex or age directly within each group by correlating normalized alpha values with sex/age, using Spearman rank correlation.

To assess difference in Beta diversity we fit a similar model as above, now correcting for sex, age and technical covariates for each individual:

$$beta\_norm \sim Sex1 + Sex2 + Age1 + Age2 + Technical\ covariates1$$
$$+ Technical\ covariates2 + Group$$

where $beta\_norm$ = beta values for each pair of individuals after inverse normal transformation, and $Group$ contains set SCZ_SCZ (both individuals from SCZ), SCZ_Control (one SCZ, one control), Control_Control (both controls).

Adjustment for pairwise comparisons for all possible disease status pairs (three comparisons) is performed using Bonferroni correction for multiple testing. We also determined a possible effect of alpha diversity on the above model by adding normalized values of alpha as a covariate to the model. In addition to our ANCOVA analysis, we performed a Permutational Multivariate Analysis of Variance (PERMANOVA) using the beta value distance matrix based on 1000 permutations (using the R function $adonis$ from the $vegan$ package).

To evaluate potential differences in microbial profiles of individuals with the different disorders (SCZ, BPD, ALS) and unaffected controls, we explored the composition and richness of the microbial communities across the groups. We observed

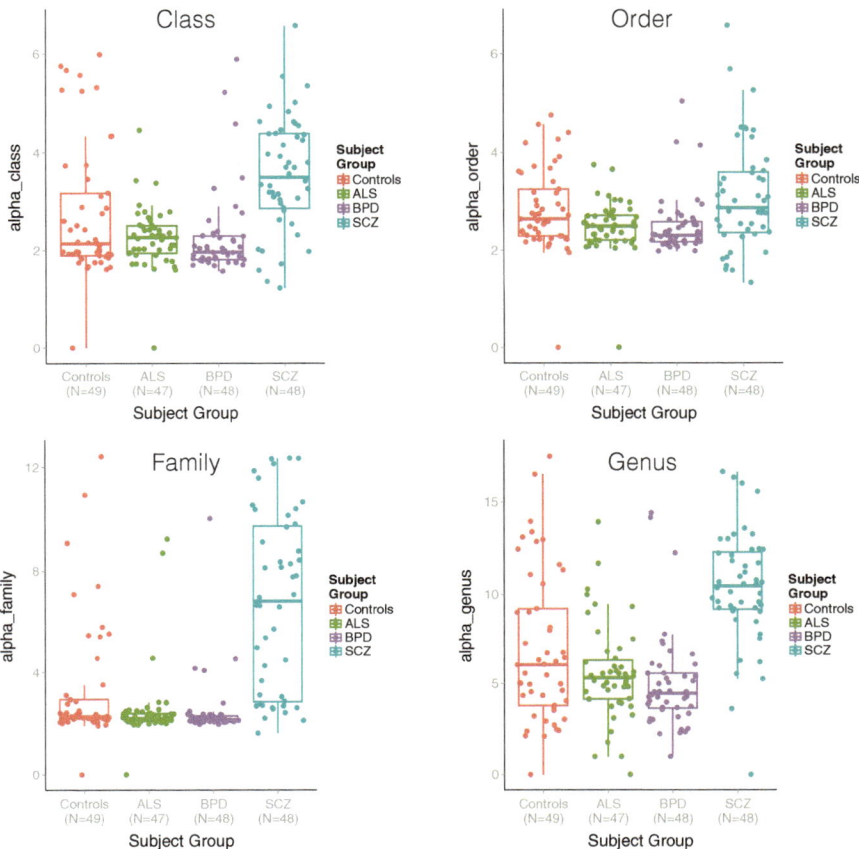

Fig. 30 Alpha diversity per sample for four subject groups (Controls, ALS, BPD, SCZ), measured using the inverse Simpson index for main taxonomic ranks (Class, Order, Family, Genus)

increased alpha diversity in schizophrenia samples compared to all other groups (*ANCOVA P* < 0.005 for all groups, Fig. 30; Tables 10 and 11, *Bonferroni* correction). These differences are corrected for covariates and are independent of potential confounders, such as experimenter and RNA extraction run (Figs. 28 and 29), and they are not the consequence of a different number of reads being detected as microbial in schizophrenia samples. No significant differences were observed between the three remaining groups (BPD, ALS, Controls). In our sample, alpha diversity was found to be a significant predictor of schizophrenia status and explained 5.0% of the variation as measured by reduction in Nagelkerke's R^2 from logistic regression. We observe no correlation between polygenic risk scores (Schizophrenia Working Group of the Psychiatric Genomics Consortium, 2014) and alpha diversity in our schizophrenia sample (n = 32, Kendall's tau = 0.008, P = 0.96). We also did not observe differences in alpha diversity between sexes or across ages, nor are our results driven by

Table 10 Alpha diversity

Disease status	Control	SCZ	BPD	ALS
Control	–	**0.0032**	1	1
SCZ	–	–	**0.0003**	**0.0128**
BPD	–	–	–	1
ALS	–	–	–	–

P-values of differences in alpha diversity across group by ANCOVA including confounding factors of sex and age, and technical covariates (RNA INtegrity (RIN) value, Plate, Flow cell lane and RNA concentration) using normalized values of alpha after Bonferroni correction for multiple testing. *P*-values that survive correction for multiple testing are marked bold

Table 11 Microbial diversity measures

Disease status	Control	SCZ	BPD	ALS
N	49	48	48	47
Alpha diversity, mean (SD)	1.77 (0.74)	2.50 (0.79)	1.55 (0.66)	1.65 (0.86)
Beta diversity, mean (SD)	0.43 (0.21)	0.50 (0.14)	0.31 (0.17)	0.38 (0.22)
Gamma diversity (per group)	20	23	18	16

the relatively younger schizophrenia cohort. Alpha diversity at other main taxonomic ranks yields a similar pattern of increased diversity in schizophrenia (Fig. 30).

We repeated the analysis of alpha diversity using only younger samples (with Age < 47, the maximum age in the schizophrenia cohort, resulting in n = 107 samples), because the groups included in the schizophrenia sample had large age differences and were younger on average. Here, we obtained similar results (i.e. *ANCOVA P* < 0.007) between schizophrenia and all other groups, and we found no significant differences observed between BPD, ALS and Controls.

To investigate whether the differences in beta diversity are driven by our primary observation of increased alpha diversity, we corrected for alpha diversity in our model. While we still observed Controls_Controls < SCZ_Controls and SCZ_SCZ < SCZ_Controls at $P < 0.001$, the beta values of Controls_Controls, and SCZ_SCZ were no longer significantly different (Fig. 29). Thus, these observations are not entirely driven by the increase in microbial diversity observed in SCZ samples.

While we normalized the number of unmapped reads to 100,000 for each sample, we hypothesized that the increased alpha diversity observed in schizophrenia may be due to an increased number of microbial reads being detected by PhyloSift in these samples. However, if we add number of reads detected as a covariate to our regression analysis the results remain unchanged (*ANCOVA P* < 0.005 for all groups with SCZ, and no significant differences were observed between the groups BPD, ALS and Controls). The average number of reads detected by PhyloSift are: SCZ (1049 ± 458), BPD (1375 ± 295), ALS (1289 ± 494) and Controls (1226 ± 478).

The increased diversity observed in schizophrenia patients may be due to specific phyla characteristic to schizophrenia, or due to a more general increased microbial

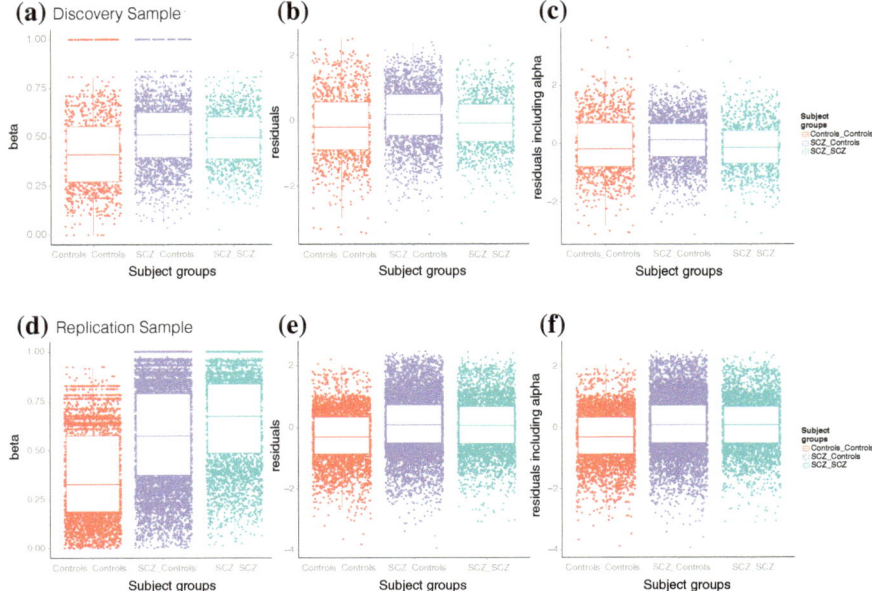

Fig. 31 Beta diversity measured using Bray-Curtis beta diversity metric across pairs of samples from the SCZ (N = 48) and Controls (N = 49), resulting in three subject groups: SCZ_Controls, SCZ_SCZ, Controls_Controls. Figures **a** shows the raw beta values. Figures **b** and **c** show beta diversity residuals after correcting for covariates (sex and age, RNA INtegrity (RIN) value, Plate, Flow cell lane and RNA concentration) (**b**), as well as alpha levels (**c**), respectively. Figures **d, e** and **f** are analogous plots for the replication study of schizophrenia cases (N = 91) and controls (N = 88)

diversity in people affected by the disease. To investigate this, we compared diversity across individuals within the schizophrenia group to control samples. We compared beta diversity across pairs of samples with schizophrenia and controls, resulting in three subject groups: SCZ_Controls, SCZ_SCZ and Controls_Controls. The lowest diversity was observed in the Controls_Controls group (0.43 ± 0.21), followed by SCZ_SCZ (0.50 ± 0.14), and the highest beta diversity values were observed for SCZ_Controls (0.51 ± 0.17) ($P < 0.05$ for each comparison, by ANCOVA after correcting for three tests). This result was confirmed by permanova ($P < 0.001$) based on 1000 permutations. Thus, the observed increased alpha diversity in schizophrenia is not caused by a particular microbial profile, but most likely represents a nonspecific overall increased microbial burden (see also Fig. 31).

In addition to measuring individual microbial diversity (alpha), and diversity between individuals (beta), we measured the total richness of the microbiome by the total number of distinct taxa of the microbiome community observed within an entire subject group [gamma diversity (Jost, 2007)]. We observed that all 23 distinct phyla are observed in schizophrenia: gamma (SCZ) = 23 compared to gamma (Controls) = 20, gamma (ALS) = 16 and gamma (BPD) = 18.

We complement the reference-based taxonomic analysis with a reference independent analysis. We utilized EMDeBruijn (https://github.com/dkoslicki/EMDeBruijn),
a reference-free approach capable of quantifying differences in microbiome composition between the samples. EMDeBruijn compresses the k-mer counts of two given
samples onto de Bruijn graphs and then measures the minimal cost of transforming
one of these graphs into the other (in terms of how many k-mers moved how far). This
direct comparison of samples allows one to circumvent the many issues involved with
selecting a phylogenetic classification algorithm, choosing which training database
to use and deciding how to compare two classifications.

Other reference-free comparison metrics have previously been used (such as treating k-mer frequencies as vectors in $\mathbb{R}^n$ and then using the Euclidean distance, Jensen-
Shannon divergence, Kullback-Liebler divergence, cosine similarity, etc.). However,
treating k-mer frequencies as vectors in $\mathbb{R}^n$ ignores the dependencies induced by the
amount of overlap between two given k-mers. Instead of Euclidean space, EMDe-
Bruijn considers k-mer frequencies as existing on an underlying de Bruijn graph, a
structure that naturally takes into consideration such overlap-induced dependencies.

Fixing a k-mer size, we first formed the undirected de Bruijn graph, with vertices
given by k-mers, and an edge between two k-mers if the first (or last) $k-1$ nucleotides
of one k-mer overlaps with the last (or first) $k-1$ nucleotides of the other k-mer. Let
$d(\cdot, \cdot)$ represent the resulting graph distance. Then, given two metagenomic samples,
S_1 and S_2, we let the frequencies of k-mer be given by $freq_k(S_1)$ and $freq_k(S_2)$
respectively. These frequencies are thought of as weights on the vertices of the de
Bruijn graph. Next, to represent the transformation of one set of weights into the
other, we used the term flow (or coupling) which is any real-valued matrix γ with
rows and columns indexed by k-mers, such that the row sums equals $freq_k(S_1)$ and
the column sums equal $freq_k(S_2)$. A flow represents how much weight was moved
where. There are infinitely many flows possible, but we chose the most efficient flow,
which is defined to be the one that minimizes the total cost (in terms of weight times
distance). This leads to the definition of the EMDeBruijn metric $EMD_k(S_1, S_1)$:

$$EMD_k(S_1, S_1) := \min_{\gamma} \sum_{x,y\text{kmers}} \gamma(x, y) * d(x, y)$$

Hence, the EMDeBruijn metric measures the minimal cost of transforming one
sample's k-mer frequency vector into the other sample's k-mer frequency vector,
when allowable transformations are restricted to moves along edges of the de Bruijn
graph. To compute this quantity, we used the FastEMD implementation of the Earth
Mover's Distance since the graph metric $d(\cdot, \cdot)$ is naturally thresholded. We found
that a good trade-off between algorithmic run-time and effectiveness of the resulting
metric was to use the k-mer size of $k = 6$.

To determine the variation explained by EMDeBruin principal components, we
adopted a similar approach as described above and fit the following logistic regression
model:

$$SCZ \sim Sex + Age + Technical\ covariates + PC1 + PC2 + PC3$$

where PCi denotes the ith EMdeBruin principal component.

To determine overlap between the results from PhyloSift and EMdeBruin, we correlated principal components of EMdeBruin and PhyloSift by Spearman rank correlation, including all samples. We complemented reference-based methods (PhyloSift and MetaPhlAn) with EMDeBruijn, a reference-independent method. EMDeBruin distances measured between samples correlated significantly with beta diversity (Spearman rank $P < 2.2e-16$, rho $= 0.37$, including SCZ and Controls). Also, EMDeBruijn PCs correlated with principal components obtained from edge PCA based on the PhyloSift taxonomic classification (correlation between EMDeBruijn PC1, and PhyloSift PC1 is $P = 1.824e-09$; Spearman rank correlation is rho $= -0.42$; see also Fig. 32). After correcting covariates, the first three EMDeBruijn PCs are significant predictors of schizophrenia status, and jointly explained 7.1% of the variance ($P < 0.05$ for each PC).

For beta diversity, the pattern we observed in the replication sample slightly diverged from the results obtained from our discovery cohort: while Controls_Controls still has the lowest average beta diversity, we observed increased beta diversity in SCZ_SCZ group versus SCZ_Controls ($P < 0.0001$, Fig. 31). One potential explanation for this discrepancy is that beta diversity in the replication sample was computed at the genus rather than phylum level, making slight mismatches between individuals more likely, and distances between samples harder to compute based on present microbial taxa. This is expected to be more likely if both samples have a large microbial diversity. In relation to this, contrary to what we observed in the discovery sample, we did not observe a correlation between EMDeBruijn distances and Beta diversity in this sample.

However, as in our discovery sample, EMDeBruijn PCs significantly correlated with principal components that were obtained from edge PCA based on the MetaPhlAn taxonomic classification (Correlation between EMDeBruijn PC1, and MetaPhlAn PC1 is $P = 6.091e-06$, rho $= -0.32$ Spearman rank correlation, see also Fig. 32). Finally, as in our discovery sample, the first three EMDeBruijn principal components adjusted for covariates were significant predictors of status and together explain 7.8% of the variance.

In addition to a global difference between schizophrenia and the other groups, we also investigated whether there are particular individual phyla contributing to the differences between schizophrenia and other groups. There are two phyla detected more often in schizophrenia cases versus all the other groups: Planctomycetes, observed in 20 SCZ cases compared to 3(ALS) 2(BPD) 5(Controls) ($P = 0.0002$ Fisher's exact for four groups, Bonferroni corrected for 23 tests $P = 0.0057$) and Thermotogae, observed in 20 SCZ cases compared to 6 ALS, 3 BPD and 6 Controls ($P = 0.0006$ Fisher's exact, corrected $P = 0.014$). No outliers were observed for the other groups (see Table 11).

We performed a replication experiment in an independent case-control sample: schizophrenia (SCZ n $= 91$) and healthy controls (Controls n $= 88$) (see Table 6d).

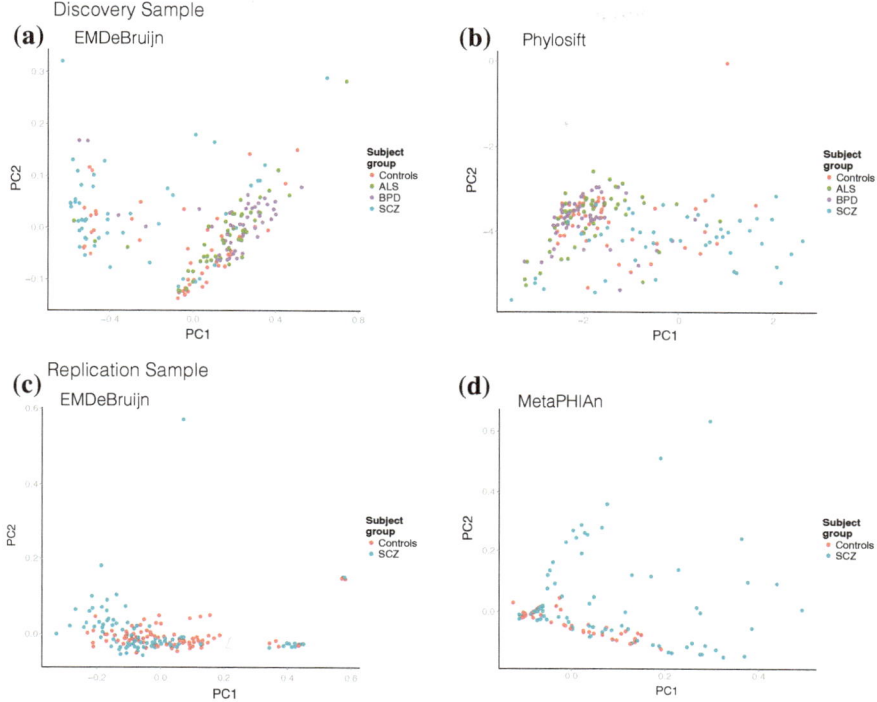

Fig. 32 Principal component analysis of microbial reads. All samples were analyzed by dimensionality-reduction methods to estimate trends in community composition with respect to disorder status. Discovery sample: **a** PCoA applied to a similarity matrix. Similarities between samples were calculated using the EMDeBruijn metric. **b** Edge PCA applied to a matrix where columns correspond to edges in the reference phylogeny, rows correspond to each sample. Each entry is the difference in placed sequence probability masses on either side of that edge calculated by Guppy using the PhyloSift output. Replication sample: **c** PCoA applied to a similarity matrix. Similarities between samples were calculated using the EMDeBruijn metric. **d** PCoA applied to a similarity matrix. Similarities between samples were calculated using Jensen-Shannon divergence on the normalized genus level reconstructions obtained by MetaPhlAn

MetaPhlAn was able to assign 5174 reads ($0.089 \pm 0.039\%$, mean $\pm$ standard deviation) on average to the bacterial gene families.

Schizophrenia samples showed increased alpha diversity on genus level (2.73 ± 0.77 for cases, versus 2.32 ± 0.57 for controls, corrected $P = 0.003$, Fig. 3b) and explained 2.5% of variance as measured by reduction in Nagelkerke R^2, thus replicating our main finding of increased diversity in schizophrenia. While our original analysis was performed on the phylum level, in our discovery sample we observe a similar increase of diversity at the genus level (Fig. 30). Similar to our discovery cohort, we observed no significant correlation between alpha diversity and age or differences across gender. Beta diversity and EMDeBruijn analyses also show similar, though not identical, patterns of nonspecific increased diversity in schizophrenia samples.

To determine a correlation between genetic risk for schizophrenia and alpha diversity, we compared alpha diversity to the polygenic risk score for schizophrenia. The polygenic risk score represents the cumulative genetic load of disease risk alleles and is defined as the sum of trait-associated alleles across many genetic loci, weighted by effect sizes estimated from a genome-wide association study. We based our scores on the most recent genome wide association study (Schizophrenia Working Group of the Psychiatric Genomics Consortium, 2014) with our samples removed (Ripke et al., 2013), and used a P-value cut-off of $P < 0.05$. For a total of 32 schizophrenia cases, we had both polygenic risk score and alpha diversity measures available, and we performed a Spearman rank correlation. We obtained similar results using different P-value cutoffs to determine the polygenic risk score.

We assessed DNA methylation data from 65 controls taken from our replication sample, and we compared methylation-derived blood cell proportions estimated using Houseman's estimation method (Houseman et al., 2012; Aryee et al., 2014) to alpha diversity after adjusting for age, gender, RIN and all technical parameters. We tested whether alpha diversity levels are associated with cell type abundance estimates. DNA methylation profiles of heterogeneous tissue types reflect variability in underlying cellular composition (Simpson, 1949; Whittaker, 1972). Recent studies, using flow-sorted cell populations, identified CpG sites discriminatory for distinct cell populations and developed sophisticated methods to estimate blood cell proportions from DNA methylation data derived from whole blood (Segata et al., 2012; Simpson, 1949; Whittaker, 1972). We used these methods to investigate a potential link between microbial diversity and the immune system.

In a control cohort of 220 individuals, blood-based genome-wide methylation data was collected using the Infinium HumanMethylation450 BeadChip. We used the epigenetic clock software (Horvath, 2013) with normalization to estimate cell abundance measures. Briefly, this software uses Houseman's estimation method (Houseman et al., 2012; Aryee et al., 2014) to estimate monocytes, granulocytes, CD8 T, CD4 T, natural killer and B cells. In addition, it predicts abundance measures for plasmablasts (i.e., immature plasma cells), CD8.naive, CD4.naive and CD8pCD28nCD45RAn cells (i.e., differentiated CD8 T cells), all based on a penalized elastic net regression model (Horvath, 2013; Horvath & Levine, 2015).

Quality control of the DNA methylation data was performed as follows. CpG sites with bead counts less than 5 or a detection p-value greater than 0.01 in more than 5% of samples were removed using the pfilter function in the wateRmelon package in R. In addition, samples having more than 5% of CpG sites with a detection p-value greater than 0.01 or having gender discrepancies were excluded from further analyses. Next, we removed CpG sites with probes containing known SNPs (EUR, MAF > 0.01) and probes that are cross-reactive, i.e. non-specific (Price et al., 2013, Chen et al., 2013). Data was background corrected using the danen function in R (wateRmelon package) and beta values were extracted for further analyses.

We investigated the relationship between microbiome diversity and the immune system as follows. From a cohort of n = 220 controls for which methylation-derived

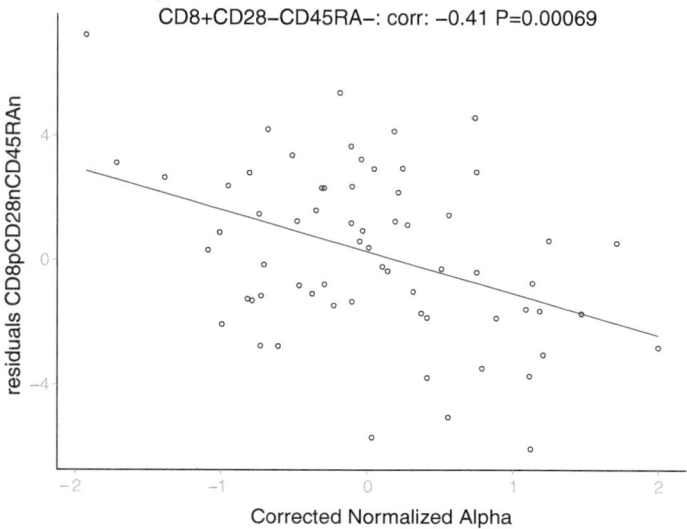

Fig. 33 Correlation between the proportion of CD8+ CD28- CD45RA-cells and corrected alpha levels at the genus level after correcting for all other cell-count estimates and technical covariates based on 65 control samples from the replication cohort

cell proportions were available, we first obtained residuals for each cell-type using the following model:

$$Proportion\_cell\_type \sim Sex + Age + Beadchip + Beadchip\ position + dataset.$$

In addition, we used residuals from the above-described regression on alpha diversity using our full replication cohort. Using all samples with both alpha levels and methylation-based cell abundance measures available (a total of n = 65), we next fitted a linear regression model with alpha diversity residuals as a response variable and all blood cell proportion residuals as independent variables. Each independent variable was analyzed as it was put in the model last to account for correlations among cell proportions. We therefore model the relationship between alpha diversity and individual cell types while adjusting for all other cell types.

We hypothesized that differences in microbial diversity may be linked to whole blood cell type composition. Our analysis shows that the proportion of one cell type, CD8$^+$ CD28$^-$ CD45RA$^-$ cells, is significantly negatively correlated with alpha diversity after correction for all other cell-count estimates as estimated from whole blood DNA methylation data (correlation = -0.41, $P = 7.3e-4$, n = 65 Controls from the Replication study, Fig. 33; Table 13). These cells are T cells that lack CD8$^+$ naïve cell markers CD28 and CD45RA and are thought to represent a subpopulation of CD8$^+$ memory T cells (Horvath & Levine, 2015; Koch et al., 2008). We observed that low alpha diversity correlates with high levels of cell abundance of this population of T cells (Table 12).

Table 12 Cell count estimates and alpha diversity

Cell Type	Method	t-value	p-value
Monocytes	Houseman	−0.33	0.74
Granulocytes	Houseman	−0.26	0.79
CD8 T cells	Houseman	0.14	0.89
CD4 T cells	Houseman	−0.88	0.38
B cells	Houseman	0.41	0.68
NK cells	Houseman	−0.5	0.62
CD8.naive	Horvath	−1.05	0.3
CD4.naive	Horvath	−1.5	0.14
Plasmablast	Horvath	0.55	0.58
CD8pCD28nCD45RAn	Horvath	−3.58	**0.00073**

Correlation of normalized and corrected alpha diversity values with cell-count estimates after correction for all other cell-count estimates, based on 73 controls from the replication sample. *P*-values that survive correction for multiple testing are marked bold

We demonstrated the prevalence and importance of the microbiome information hidden in RNA-Sequencing datasets. These previously unexplored microbial reads reveal valuable underlying clinical information that can be used for potential molecular markers. Also, the immune cell type composition of the tissue can be inferred from the RNA-Seq data to elucidate microbial diversity. Thus, it is essential to mine the hidden treasure of RNA-Sequencing to reveal the impact of immune system in human beyond the scope of microbiome.

A key function of the adaptive immune system is to mount protective memory responses to a given antigen. B cells recognize their specific antigens through their surface antigen receptors (immunoglobulins, Ig), which are unique to each cell and its progeny. A typical Ig repertoire is composed of one immunoglobulin heavy chain (IGH) and two light chains, kappa, and lambda (IGK and IGL). Igs are diversified through somatic recombination, a process that randomly combines variable (V), diversity (D), and joining (J) gene segments, and inserts or deletes non-templated bases at the recombination junctions (Georgiou et al., 2014) (Fig. 34a). The resulting DNA sequences are then translated into antigen receptor proteins. This process allows for an astonishing diversity of the Ig repertoire (i.e., the collection of antigen receptors of a given individual), with $>10^{13}$ theoretically possible distinct Ig receptors (Georgiou et al., 2014). This diversity is key for the immune system to confer protection against a wide variety of potential pathogens (Freeman, Warren, Webb, Nelson, & Holt, 2009). In addition, upon activation of a B cell, somatic hypermutation further diversifies Ig in their variable region. These changes are mostly single-base substitutions occurring at extremely high rates (10^{-5}–10^{-3} mutations per base pair per generation) (Rajewsky, Forster, & Cumano, 1987). Isotype switching is another mechanism that contributes to B-cell functional diversity. Here, antigen specificity

Table 13 Microbial phyla prevalence across individuals and disorders

Microbial phyla	Controls	Amyotrophic lateral sclerosis (ALS)	Bipolar disorder (BPD)	Schizophrenia (SCZ)	Total	P-value Fisher	P-value corrected
Fusobacterium	0	0	0	1	1	1.0000	1.0000
Elusimicrobia	1	1	0	1	3	1.0000	1.0000
Acidobacteria	2	0	0	1	3	0.6209	1.0000
Tenericutes	0	1	0	2	3	1.0000	1.0000
Deinococcus-Thermus	1	0	2	1	4	0.9561	1.0000
Synergistetes	0	2	1	2	5	0.9561	1.0000
Aquificae	1	0	0	4	5	0.1826	1.0000
Nitrospirae	3	0	1	2	6	0.5012	1.0000
Spirochaetes	4	0	1	1	6	0.4851	1.0000
Chlamydiae	1	2	1	4	8	0.9004	1.0000
Chloroflexi	2	2	1	4	9	0.8710	1.0000
Verrucomicrobia	3	1	4	3	11	0.8516	1.0000
Deferribacteraceae	3	0	2	9	14	0.0092	0.2126
Thaumarchaeota	6	4	4	4	18	0.9678	1.0000
Bacteroidetes	3	5	2	15	25	0.0888	1.0000
Planctomycetes	5	3	2	20	30	0.0002	**0.0057**
Thermotogae	6	3	6	20	35	0.0006	**0.0141**
Euryarchaeota	11	16	15	6	48	0.0669	1.0000
Crenarchaeota	18	16	17	4	55	0.0215	0.4955
Actinobacteria	15	10	12	19	56	0.2420	1.0000
Cyanobacteria[a]	28	19	15	32	94	0.0043	0.0991
Firmicutes[a]	36	29	32	43	140	0.0130	0.2988
Proteobacteria[a]	48	46	48	48	190	0.8853	1.0000

P-values that survive correction for multiple testing are marked bold. Phyla marked with a [a] are present in at least 1/4th of the samples in each subject group

remains unchanged while the heavy chain VDJ regions join with different constant (C) regions, such as IgG, IgA, or IgE isotypes, and alter the immunological properties of a BCR. The pairing of heavy and light occurring in polyclonally activated B cells chains is another mechanism to increase the Ig diversity.

High-throughput technologies enable unprecedented accuracy when profiling the Ig repertoires. Commonly used assay-based approaches provide a detailed view of the adaptive immune system with deep sequencing of amplified DNA or RNA from the variable region of the Ig locus (BCR-Seq) (Benichou, Ben-Hamo, Louzoun, & Efroni, 2012; DeWitt et al., 2016, Putintseva et al., 2013). Those technologies are usually restricted to one chain, with the majority of studies focusing on the heavy chain of Ig repertoire. Recent studies (Freeman et al., 2009) successfully applied assay-based

Table 14 Concordance of targeted BCR-Seq and non-specific RNA-Seq performed on 13 tumor biopsies from Burkitt lymphoma

ID	Major clonotype frequency (BCR-Seq) [%]	Minor clonotype frequency (BCR-Seq) [%]	Portion of IGH repertoire captured by ImRep (%)	Portion of IGH repertoire captured by MiXCR (%)	Major clonotype frequency (ImReP) [%]	Minor clonotype frequency (ImRep) [%]	Major clonotype frequency (MiXCR) [%]	Minor clonotype frequency (MiXCR) [%]	Number of BCR-SEQ-confirmed clonotypes	Number of ImReP-derived clonotypes (RNA-Seq)	Number of MiXCR-derived clonotypes (RNA-Seq)
009-0192	0.64	0.16	7.48	7.96	0.64	0.16	0.64	0.16	583	34	32
009-0184	1.11	1.11	1.11	0.00	1.11	1.11	0.00	0.00	90	1	0
009-0148	1.45	0.15	0.00	0.00	0.00	0.00	0.00	0.00	671	0	0
009-0171	2.94	1.47	2.94	0.00	2.94	2.94	0.00	0.00	61	1	0
009-0203	3.33	0.83	0.00	0.00	0.00	0.00	0.00	0.00	117	0	0
009-0186	16.67	0.67	28.67	20.00	16.67	0.67	16.67	0.67	113	15	5
009-0174	62.54	0.34	62.54	0.00	62.54	62.54	0.00	0.00	55	1	0
009-0109	96.66	0.01	96.70	0.01	96.66	0.01	0.01	0.01	223	4	1
009-0249	97.68	0.00	97.68	97.68	97.68	97.68	97.68	97.68	379	1	1
009-0112	97.75	0.00	97.75	97.75	97.75	97.75	97.75	97.75	201	1	1
009-0122	98.16	0.00	98.17	98.16	98.16	0.00	98.16	0.00	119	3	2
009-0202	99.72	0.00	99.72	99.75	99.72	99.72	99.72	99.72	50	1	2
009-0103	99.75	0.03	99.75	99.75	99.75	99.75	99.75	99.75	9	1	1

(a) Schematic representation of human Ig repertoires

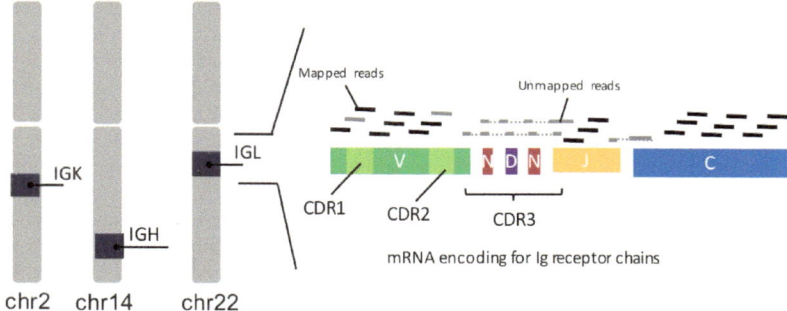

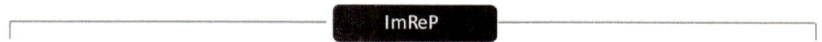

(b) Alignment-free detection of reads containing full length CDR3s and overlapping V and J genes

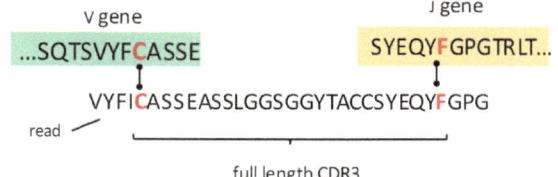

(c) Match reads containing partial CDR3s and overlapping only V or J genes

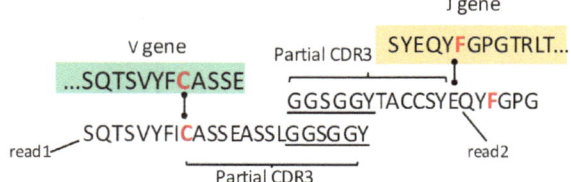

(d) Correcting PCR and sequencing errors via CAST clustering

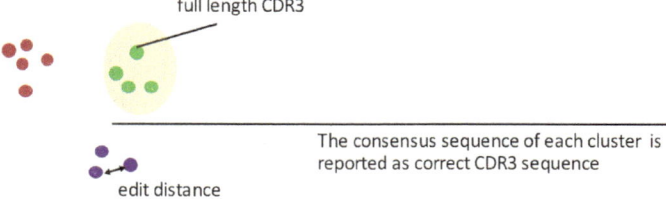

◄**Fig. 34** Overview of ImReP. **a** Schematic representation of human Ig receptor repertoire. Ig repertoire consists of three immunoglobulin loci (red color, Immunoglobulin heavy locus (IGH); Immunoglobulin kappa locus (IGK); Immunoglobulin lambda locus (IGL). Ig receptors contain multiple variable (V, green color), diversity (D, present only in IGH, violet color), joining (J, yellow color) and constant (C, blue color) gene segments. V(D)J gene segments are randomly jointed and non-templated bases (N, dark red color) are inserted at the recombination junctions. The resulting spliced Ig repertoire transcript incorporates the C segment and is translated into the antigen receptor proteins. RNA-Seq reads are derived from the rearranged immunoglobulin Ig loci. Reads entirely aligned to IG genes are inferred from mapped reads (black color). Reads with extensive somatic hypermutations and reads spanning the V(D)J recombination are inferred from the unmapped reads (grey color). Complementarity determining region 3 (CDR3) is the most variable region of the three CDR regions and is used to identify Ig receptor clonotypes—a group of clones with identical CDR3 amino acid sequences. **b** Alignment-free detection of reads containing full-length CDR3s and simultaneously overlapping V and J genes. Receptor derived reads spanning V(D)J recombinations are identified from unmapped reads and assembled into the CDR3 sequences. We first scan the amino acid sequences of the read to determines putative CDR3 sequences fully contained inside the read. The CDR3 sequence is a sequence starting with cysteine (C) and ending with (F) (IGK and IGL) or tryptophan (W) (for IGH). Reads with putative CDR3s are further examined to simultaneously overlap V and J gene segments. The alignment between the read and V and J genes is found by matching the prefix and suffix of the read to match the suffix of V and prefix of J genes, respectively. **c** Match reads containing partial CDR3s and overlapping only V or J genes. In case a read contains a partial CDR3 sequence and overlaps with only the V or J gene, we perform the second stage of ImReP. During this stage we match reads originated from the same CDR3 based on 15 nucleotides overlap. **d** Correcting PCR and sequencing errors via CAST clustering. We further correct PCR and sequencing errors in the assembled CDR3s. ImReP clusters assembled CDR3 into a set of clusters via CAST algorithm. The consensus sequence of each cluster is reported as correct CDR3 sequence

approaches to characterize the immune repertoire of the peripheral blood. However, little is known about the immunological repertoires of other human tissues, including barrier tissues like skin and mucosae. Studies involving assay-based protocols usually have small sample sizes, thus limiting analysis of *intra*-individual variation of immunological receptors across diverse human tissues.

RNA Sequencing (RNA-Seq) traditionally uses the reads mapped onto human genome references to study the transcriptional landscape of both single cells and entire cellular populations. In contrast to assay-based protocols that produce reads from the amplified variable region of Ig locus, RNA-Seq is able to capture the entire cellular population of the sample, including B cells. However, due to the repetitive nature of the Ig locus, as well as the extreme level of diversity in Ig transcripts, most mapping tools are ill-equipped to handle *Ig* sequences. RNA-Seq was successfully used for analysis of highly clonal leukemic repertoires with high relative quantities of *Ig* transcripts (DeWitt et al., 2016). Despite this, Ig transcripts often occur in sufficient numbers within the transcriptome of many tissues to characterize their respective Ig repertoires (Blachly et al., 2015). A number of methods (Bolotin et al., 2015; Li et al., 2016; Stubbington et al., 2016) were designed to assemble Ig and T cell receptor repertoires and have been applied across various public RNA-Seq datasets. Existing methods that are capable of assembling Ig repertoires from bulk RNA-Seq data produce low-accuracy results (f-score < 0.2) relative to ImReP (Fig. 38a).

We developed ImReP, a novel alignment-free computational method for rapid and accurate profiling of the Ig repertoire from regular RNA-Seq data. We applied it to 8555 samples across 544 individuals from 53 tissues obtained from Genotype-Tissue Expression study (GTEx v6) (GTEx Consortium et al., 2017). The data was derived from 38 solid organ tissues, 11 brain subregions, whole blood, and three cell lines. ImReP is able to efficiently extract Ig-derived reads from the RNA-Seq data and accurately assemble the complementarity determining regions 3 (CDR3s). CDR3 are the most variable regions of the Ig receptors determining the antigen specificity. Using ImReP, we created a systematic atlas of Ig sequences across a broad range of tissue types, most of which were not previously studied for Ig repertoires. We also examined the compositional similarities of clonal populations between the tissues to track the flow of Ig clonotypes across immune-related tissues, including secondary lymphoid and organs that encompass mucosal, exocrine, and endocrine sites. Our proposed approach is not superior in comparison to targeted BCR-Seq; rather, it provides a useful tool for mining large-scale RNA-Seq datasets for the study of *Ig* receptor repertoires.

A number of tools have been developed to reconstruct the *Ig* receptor repertoire. Repertoire analysis from RNA-Seq data typically starts with mapping the reads to the germline V, D, and J genes obtained from the International ImMunoGeneTics (IMGT) database (Lefranc et al., 2015). There are three possible read mapping scenarios: (1) the read is entirely mapped to the V gene; (2) the read is entirely mapped to the J gene; (3) the read is partially mapped to the V and J genes simultaneously. Existing methods consider only reads from category (3). These methods use different underlying algorithms to map reads to germline genes. For example, MiXCR (Bolotin et al., 2015) relies on an in-house alignment procedure, IgBlast (Ye et al., 2013) utilizes BLAST with an optimized set of parameters, and IMSEQ (Kuchenbecker et al., 2015) uses in-house pairwise alignment between the read sequence and the germline V and J segment sequences.

Following the alignment, MiXCR performs overlapping of previously aligned reads into contigs. The resulting contigs are re-aligned to the V, D, and J genes to verify that the significant portion of non-template N insertions is covered. In contrast to MiXCR, which simultaneously aligns reads to both V, D, and J genes, IgBlast separately aligns the query read to the databases comprised of V, D, and J genes. IgBlast uses a specific sequence to separately align; first, the program finds the best V gene hit. Then, IgBlast masks the aligned read region and performs an alignment to the J gene database. (In the event of a heavy chain, IgBlast also queries the D gene database for the best hit.) The software checks that each component in the obtained V(D)J rearrangement originates from the same locus.

All methods use the definition of CDR3 to determine the boundaries of CDR3 sequence in each of the read. The last step in repertoire analysis is to correct the assembled clones for PCR and sequencing errors. In order to correct such errors, MiXCR and IMSEQ cluster the assembled clones and report a consensus sequence per cluster. IgBlast skips the error correction step and directly outputs inferred clones.

Most methods use alignment or assembly to infer CDR3s and align reads to V and J genes. In contrast, the ImReP procedure provides a match the read prefix

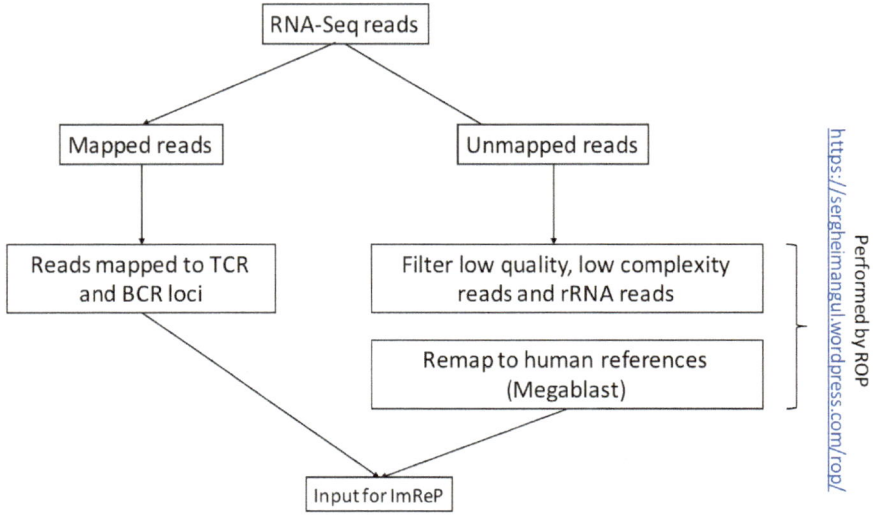

Fig. 35 Schematic on how to select the candidate receptor-derived reads from RNA-Seq reads, which are the input for ImReP

and read suffix to the suffix of V and prefix of J genes, respectively, without using alignment. Avoiding alignment allows ImReP to significantly decrease running time and computational resources required to run the package. Average CPU time reported for ImReP is 44 min; this runtime is substantially shorter than the 10 h required for MiXCR. On average per sample, ImReP consumed 3G of CPU while MiXCR required 10G of CPU.

We applied ImReP to 0.6 trillion RNA-Seq reads (92 Tbp) from 8555 samples to assemble CDR3 sequences IG receptors (Table 15). The RNA-Seq data was generated by the Genotype-Tissue Expression Consortium (GTEx v6). First, we mapped RNA-Seq reads to the human reference genome using a short-read aligner (performed by GTEx consortium[11]) (Fig. 34). Next, we identify reads spanning the V(D)J junction of the IG receptors and assemble clonotypes (a group of clones with identical CDR3 amino acid sequences). We defined the CDR3 as the sequence of amino acids, starting with the cysteine on the left of the junction and ending with phenylalanine (for IGK, and IGL) or tryptophan (for IGH) on the right of the junction. Here ImReP used 0.02 trillion high quality reads that were successfully mapped to Ig genes or were unmapped reads that failed to map to the human reference genome (Figs. 34a and 35).

ImReP is a two-stage alignment-free approach to assemble CDR3 sequences and detect corresponding V(D)J recombinations (Fig. 34b). In the first stage, we prepare the candidate receptor reads from mapped and unmapped RNA-Seq reads (Fig. 35). We merge partially mapped reads from BCR loci and unmapped reads into a set of candidate receptor reads, which serves as an input for ImReP. We scan the *amino acid* sequences of the read and determine the putative CDR3 as a substring of the

Table 15 Data overview

	Body site	Tissue type	Histological type	Reads (millions)	No. samples	CD19+ B cells	CDR3s [IGH]	CDR3s [IGK]	CDR3s [IGL]	CPM IGH	CPM IGK	CPM IGL	IGH-derived reads	IGK-derived reads	IGL-derived reads	Diversity IGH	Diversity IGK	Diversity IGL
col#	1	2	3	4	5	6	9	10	11	13	14	15	17	18	19	21	22	23
	Spleen	Secondary lymphoid organs	Spleen	18	11	1.8	194	730	543	901	400	304	226	206	135	7.7	7.4	7.4
	Small Intestine - Terminal Ileum		Small Intestine	20	10	1.2	201	217	212	101	139	108	196	306	271	6.5	6.8	6.3
	Whole Blood	Blood associated sites	Blood	27	54	1.2	315	808	670	19	33	23	44	119	77	5.2	5.8	5.3
	Artery - Coronary		Blood Vessel	19	14	0.5	112	171	131	6	9	6	22	91	42	3.3	3.7	3.3
	Artery - Aorta		Blood Vessel	19	24	0.4	92	112	80	4	6	4	11	50	18	5.2	5.6	5.2
	Artery - Tibial		Blood Vessel	23	36	0.2	16	32	21	1	1	1	3	8	6	1.3	1.2	1.6
	Minor Salivary Gland	Mucosal, exocrine and endocrine organs	Salivary Gland	19	70	0.7	535	485	348	181	148	179	315	898	451	7.1	7.2	7.0
	Colon - Transverse		Colon	19	20	0.7	287	379	258	131	109	136	246	780	164	6.6	6.8	6.3
	Stomach		Stomach	19	20	0.6	845	139	100	44	72	33	98	237	105	5.3	5.7	5.4
	Colon - Sigmoid		Colon	20	17	0.3	646	765	389	32	56	30	67	137	96	5.8	5.9	5.2
	Lung		Lung	25	13	0.6	621	101	733	19	41	30	37	174	110	5.4	5.7	5.4
	Breast - Mammary Tissue		Breast	20	21	0.5	439	625	433	21	34	21	42	135	65	4.1	4.3	4.2
	Esophagus - Mucosa		Esophagus	19	32	0.4	503	571	373	16	30	20	27	94	44	4.4	5.0	4.6
	Thyroid		Thyroid	26	20	0.4	303	468	175	12	18	11	46	168	72	5.1	5.8	5.2
	Kidney - Cortex		Kidney	21	27	0.4	280	362	318	15	26	13	16	96	52	5.2	5.3	5.3
	Bladder		Bladder	15	11	0.4	239	341	247	19	23	16	45	114	98	5.2	5.8	5.1
	Fallopian Tube		Fallopian Tube	17	28	0.6	228	403	238	13	26	13	15	111	39	4.5	4.7	4.3
	Vagina		Vagina	18	90	0.4	168	256	161	8	14	8	23	51	26	3.8	4.1	3.7
	Cervix - Endocervix		Cervix Uteri	15	5	0.3	160	335	161	11	22	11	52	109	51	3.7	4.3	4.0
	Liver		Liver	20	13	0.4	142	365	218	7	18	11	10	33	17	4.1	5.2	4.7
	Cervix - Ectocervix		Cervix Uteri	16	5	0.4	133	314	132	8	20	8	13	49	18	3.8	4.4	3.4
	Prostate		Prostate	19	12	0.4	76	190	113	4	10	6	7	31	18	3.2	4.2	3.5
	Uterus		Uterus	19	36	0.3	37	134	71	3	8	4	4	15	7	2.8	3.8	3.4
	Ovary		Ovary	19	33	0.2	49	109	64	3	5	3	4	23	7	3.7	3.3	3.1
	Pancreas		Pancreas	20	10	0.2	42	65	62	2	3	3	3	8	3	3.7	3.8	3.2
	Pituitary		Pituitary	21	18	0.4	21	63	33	1	2	2	3	5	3	2.3	3.2	2.8
	Skin - Not Sun Exposed (Suprapubic)		Skin	21	11	0.3	20	40	28	1	2	1	2	7	4	2.1	2.8	2.1
	Skin - Sun Exposed (Lower leg)		Skin	23	27	0.3	11	23	15	0	1	1	0	4	2	1.0	1.8	1.0
	Adrenal Gland		Adrenal Gland	20	15	0.4	52	62	67	2	3	2	6	19	15	3.4	5.5	3.7
	Testis		Testis	20	10	0.3	9	21	14	0	1	1	1	3	2	1.4	2.5	1.8
	Cells - EBV-transformed lymphocytes	Cell lines	Blood	25	13	1.8	49	54	31	2	2	1	238	572	281	1.0	1.0	1.1
	Cells - Transformed fibroblasts		Skin	20	30	0.1	5	6	4	0	0	0				0.8	1.8	1.1
	Adipose - Visceral (Omentum)	Adipose and muscle organs	Adipose Tissue	20	23	0.6	108	138	100	3	8	3	14	44	21	3.6	3.9	3.5
	Esophagus - Gastroesophageal Junction		Esophagus	19	17	0.5	71	129	94	4	7	4	11	30	17	3.8	3.7	3.2
	Adipose - Subcutaneous		Adipose Tissue	23	38	0.3	23	41	25	1	3	1	2	8	3	2.3	2.8	2.4
	Heart - Atrial Appendage		Heart	23	7	0.3	18	34	21	1	3	1	2	8	3	3.8	3.8	3.1
	Esophagus - Muscularis		Esophagus	19	2	0.3	13	32	20	2	4	2	2	4	2	2.7	3.3	3.1
	Heart - Left Ventricle		Heart	26	21	0.2	9	11	15	0	1	0	1	3	1	1.4	2.4	1.8
	Muscle - Skeletal		Muscle	24	70	0.1	6	17	6	0	1	0	0	1	1	1.2	2.1	1.8

Table 15 (continued)

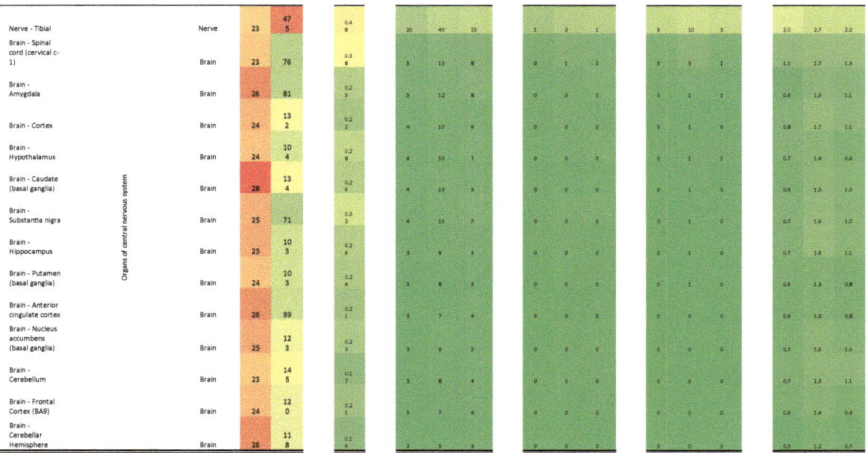

Characteristics of 8555 samples across 544 individuals from 53 body sites obtained from Genotype-Tissue Expression study (GTEx v6). The second column reports the tissue type based on the relationship to the immune system (main text). The tissues inside each tissue type group are sorted based on number of CDR3 sequences. (3) Histological type of the body site. (4) The median number of 76 × 2 bp paired-end reads per sample. (5) Number of RNA-Seq samples available via GTEx. Results for (7–40) are presented individually for immunoglobulin heavy chain (IGH), immunoglobulin kappa chain (IGK, immunoglobulin lambda chain (IGK), T cell receptor alpha chain (TCRA), T cell receptor beta chain (TCRB), T cell receptor delta chain (TCRD), and T cell receptor gamma chain (TCRG). (7–8) Median relative abundance of B or T cells within each tissue. (10–16) Median number of distinct CDR3 (clonotypes) is reported per tissue. (18–24) Median number of distinct clonotypes (CDR3) per 1 million RNA-Seq reads (CPM) is reported. (34–40) We used per sample alpha diversity (Shannon entropy) to estimate the diversity of immune repertoire. Median value per tissue is reported

read starting from *cysteine (C)* and ending with *phenylalanine (F)* (or *tryptophan (W)* for IGH). A read is decomposed in three parts: read prefix, CDR3, read suffix. The CDR3 sequence is a sequence starting with cysteine (C) and ending with (F) (IGK and IGL) or tryptophan (W) (for IGH). Reads with putative CDR3s are further examined to assess the overlap with V and J genes. Variable Ig receptor genes were imported from IMGT version: 3.1.17. We used C from the read (starting of CDR3) and C from the V gene (which usually occur as the three before the last codon in the V sequence) as an anchor to align the read prefix and V gene. Similarly, we used F (or W) from the read (end of CDR3) and F (or W) (which usually occur as the third codon in the J sequence). In the second stage, ImReP utilizes reads that contain a partial CDR3 sequence and overlap a single gene segment (V or J). Using the alignment-free procedure described above, we determined the alignment between the V or J gene and the read prefix or suffix, respectively. ImReP performs matching with a suffix tree technique. Matched reads with an overlap of at least 15 nucleotides are used to assemble full-length CDR3s. We further correct PCR and sequencing errors in the assembled CDR3s. ImReP clusters assembled CDR3 into a set of clusters via the CAST algorithm (Stubbington et al., 2016). The clustering procedure is iteratively repeated until the average inverse edit distance (Levenshtein) inside each cluster is less than the user-defined threshold (ImReP default is 0.2). The consensus sequence of each cluster is reported as the correct CDR3 sequence.

ImReP is a computational approach to assemble CDR3 sequences and detect corresponding V(D)J recombinations from B and T cell receptors. ImReP is composed of two stages. In the first stage, ImReP utilizes the reads that simultaneously overlap V and J gene segments to infer the CDR3 sequences. We define the CDR3 as the sequence of amino acids between the cysteine on the right and phenylalanine (for TCR, IGK, and IGL) or tryptophan (for IGH) on the left of the junction. We first convert the read sequences from nucleotides to amino acids. We scan the *amino acid* sequences of the read and determine the putative CDR3 as a subsequence of the read starting from cysteine (C) and ending with phenylalanine (F) (and tryptophan (W) for IGH). The reads containing the described substring are considered candidate CDR3 reads. We denoted n to be the length of the read. We denoted the coordinates of putative CDR3 string to be x and y, corresponding with the start and end of the CDR3 sequence in the read coordinates. This way each candidate CDR3 read is composed of three parts:

1. $r[0, x - 1]$ is a prefix of the read, potentially overlapping suffix of V gene. It contains the amino acids from the read, from position 0 to $x - 1$.
2. $r[x, y]$ is a substring of the read containing the putative CDR3 sequence. It contains the amino acids from the read, from position x to y.
3. $r[y + 1, n]$ is a suffix of the read potentially overlapping prefix of J gene. It contains the amino acids of the read, from position $y + 1$ to n.

The amino acid sequences of V and J genes of B cell receptors (BCR) and T cell receptors (TCR) were imported from IMGT (International ImMunoGeneTics information system). For each V gene, we identify last conserved cysteine (C) and record its position p_C. For each J gene, we identify first conserved phenylalanine (for IGK, IGL, TCRA, TCRB, TCRG, TCRD) or tryptophan (for IGH) and record its position p_F. For each V gene, we extract two substrings: $V_x = V[0, p_C - 1]$ and $V_y = [p_C + 1, nV]$. For each J gene, we record two substrings: $Jx = J[0, p_F - 1]$ and $Jy = J[p_F + 1, nJ]$, where nV and nJ are the lengths of V and J genes, respectively. Given a set of candidate CDR3 reads, we attempt to find the corresponding V and J genes. We match a substring of the read $r[0, x - 1]$ with the corresponding suffix of V_x for V genes. We also match the read $r[y + 1, n]$ with the corresponding prefix of J_x for J genes. We consider a read to match V gene if the length of $r[0, x - 1]$ is greater than 4 and the edit distance between $r[0, x - 1]$ and V_x is less than 2. We consider a read to match J gene if the length of $r[y + 1, n]$ is greater than 4 and the edit distance between $r[y + 1, n]$ and Jx is less than 2. In case a read overlaps equally (in terms of edit distance) among multiple V genes and J genes, we report all of them.

In the second stage, ImReP utilized the reads overlapping only V or J gene. Such reads contain a partial CDR3 sequence. ImReP builds a suffix tree S on the reads overlapping any of the V genes. Then, for each read j overlapping a J gene a V-gene overlapping read, v from S is determined (in case if any exists). Reads v and j are concatenated (with overlap), and the CDR3 region is extracted.

Further, ImReP uses a CAST clustering technique to correctly assemble CDR3s for PCR and sequencing errors. The output of the algorithm is the set of CDR3 partitions, and each of the partition corresponds to a clonotype. Specifically, ImReP builds a complete graph $G = (V, E, w)$, where the set of vertices V is represented by the set of assembled CDR3 sequences. The weight of the edge is determined by the inverse of the edit distance, computed between the two CDR3 sequences x and y. The CAST algorithm is executed with the following procedure. A new partition P is initialized with the max-degree node. Then, the set of "close" vertices is iteratively added to the partition, and the set of "distant" vertices are removed from the partition. A vertex v is deemed to be "close" ("distant"), if the average distance from v to the vertices from P is greater (smaller) than a user-defined threshold. The procedure is repeated until either the set of "close" or the set of "distant" vertices is empty. In such a way, the partition P is based on a max-degree node and extended with the "close" vertices. Vertices belonging to P are then removed from the graph G, and the clustering procedure is repeated until all of the vertices are assigned to a partition. Let $\{v_1, v_2, \ldots, v_i, \ldots, v_n\}$ be a partition output by the CAST algorithm. Each v_i has an associated weight equal to the count of CDR3's v_i, which was assembled during the first two stages of ImReP. We compute the weighted consensus sequence of P and output the sequence as a final clonotype. Finally, we map D genes (for IGH, TCRB, TCRG) onto assembled CDR3 sequences and infer corresponding V(D)J recombination. Starting with release v0.8, ImReP reports out of frame CDR3 sequences.

To validate the feasibility of using RNA-Seq to study the Ig receptor repertoire, we simulated RNA-Seq data as a mixture of transcriptomic reads and reads derived from IG transcripts (ratio between Ig-derived reads and transcriptomic reads was on average 1: 3600) (Fig. 36). Ig transcripts are simulated based on random recombination of V, D and J gene segments (obtained from IMGT database) (Lefranc et al., 2015) with non-template insertion at the recombination junctions (Fig. 37). We assessed the ability of ImReP to extract CDR3-derived reads from the RNA-Seq mixture by applying ImReP to a simulated RNA-Seq mixture. While our simulation approach may not completely summarize the various nuances and eccentricities of actual immune repertoires, it allows us to assess the accuracy of our tool. ImReP is able to identify 99% of CDR3-derived reads from the RNA-Seq mixture, suggesting it is a powerful tool for profiling RNA-Seq samples of immune-related tissues.

Next, we compared ImReP with other methods designed to assemble *Ig* receptor repertoires. We also investigated the sequencing depth and read length required to reliably assemble *Ig* sequences from RNA-Seq data. Our simulations suggest that both read length and sequencing depth have a major impact on precision-recall rates of CDR3 sequence assembly. ImReP is able to maintain an 80% precision rate for the majority of simulated scenarios. Average CDR3 coverage that is higher than 8 allows ImReP to archive a recall rate close to 90% for a read length above 75 bp (Fig. 38a). Increasing coverage has a positive effect on the number of assembled clonotypes achieved by ImReP.

Fig. 36 Workflow of BCR and TCR transcript simulations

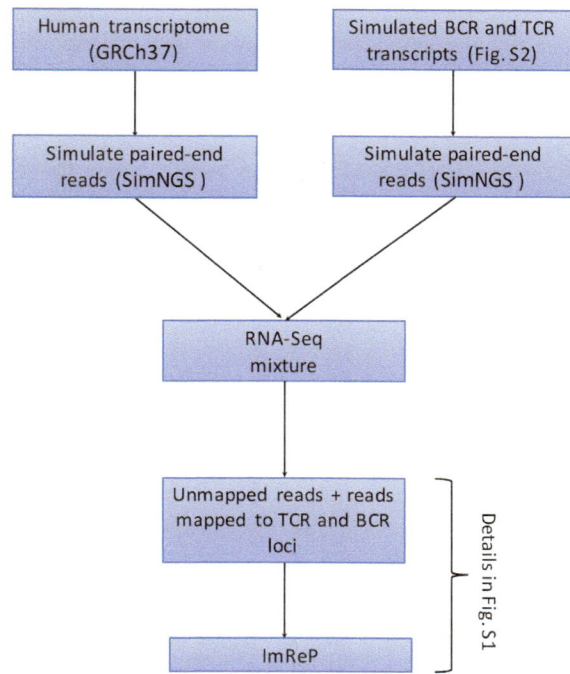

We compared the performance of ImReP to MiXCR (RNA-Seq mode) (Bolotin et al., 2015; Kuchenbecker et al., 2015), IgBlast-based pipeline (Strauli & Hernandez, 2016), and IMSEQ (Kuchenbecker et al., 2015). Except for IMSEQ, these tools were developed to assemble the hypervariable sequences from Ig receptors directly from RNA-Seq data. Another tool, iSSAKE (Warren, Nelson, & Holt, 2009), is no longer supported and was not recommended for use. Unfortunately, we obtained empty output after running V'DJer (Mose et al., 2016), and increasing coverage in the simulated data did not solve the problem. We exclude TRUST (Li et al., 2016) and TraCeR (Stubbington et al., 2016), as those methods are solely designed for T cell receptors. We supplied each of those tools with the original RNA-Seq reads as raw or mapped reads, depending on the software developers' recommendations. IMSEQ (Kuchenbecker et al., 2015) cannot be applied directly to RNA-Seq reads because they were originally designed for targeted sequencing of *Ig* receptor loci. Thus, to independently assess and compare accuracy with ImReP, we only ran IMSEQ with the simulated reads derived from BCR transcripts (Fig. 36). ImReP consistently outperformed existing methods in both recall and precision rates. The recall was defined as $TP/(TP + FN)$. Precision was defined as $TP/(TP + FP)$. We defined TP as the number of correctly assembled CDR3 sequence (based on the exact match), FN was defined as the number of true CDR3 sequence not assembled by the method, and FP was defined as the number of incorrectly assembled CDR3 sequences. On average, ImReP offers three-time superior accuracy (average f-score of ImRep was 0.78, for

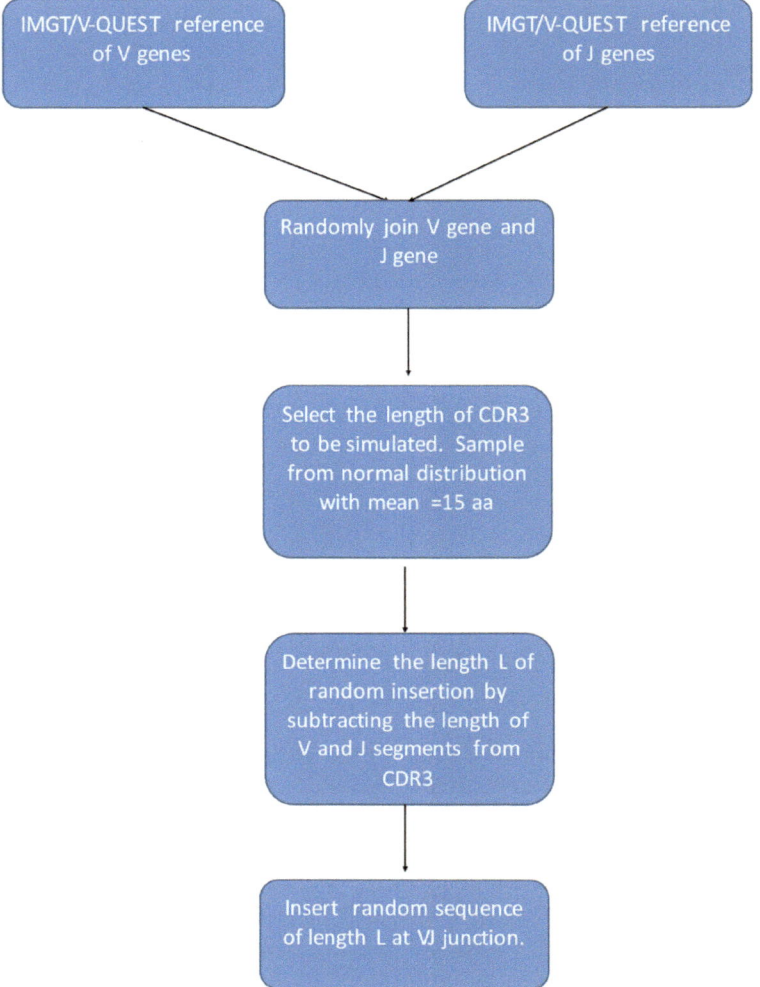

Fig. 37 Schematic on how the mixture of transcriptomic and receptor-derived reads was generated

other methods average f-score was <0.2). F-score was defined as a harmonic mean of precision and recall. Notably, ImReP was the only method with acceptable performance for 50 bp read length, reconstructing with a higher precision rate significantly and more CDR3 clonotypes than other methods.

To further demonstrate the feasibility of applying non-specific RNA Sequencing to profile *Ig* receptor repertoires, we have used 18 tumor biopsies sequenced by BCR-Seq and RNA-Seq. Biopsies were acquired from the patients with histologically confirmed Burkitt lymphoma (Lombardo et al., 2017). Per sample, 100 million paired-end RNA-Seq reads of length 50 bp were available. We first mapped the reads onto the reference human genome and transcriptome and extracted unmapped

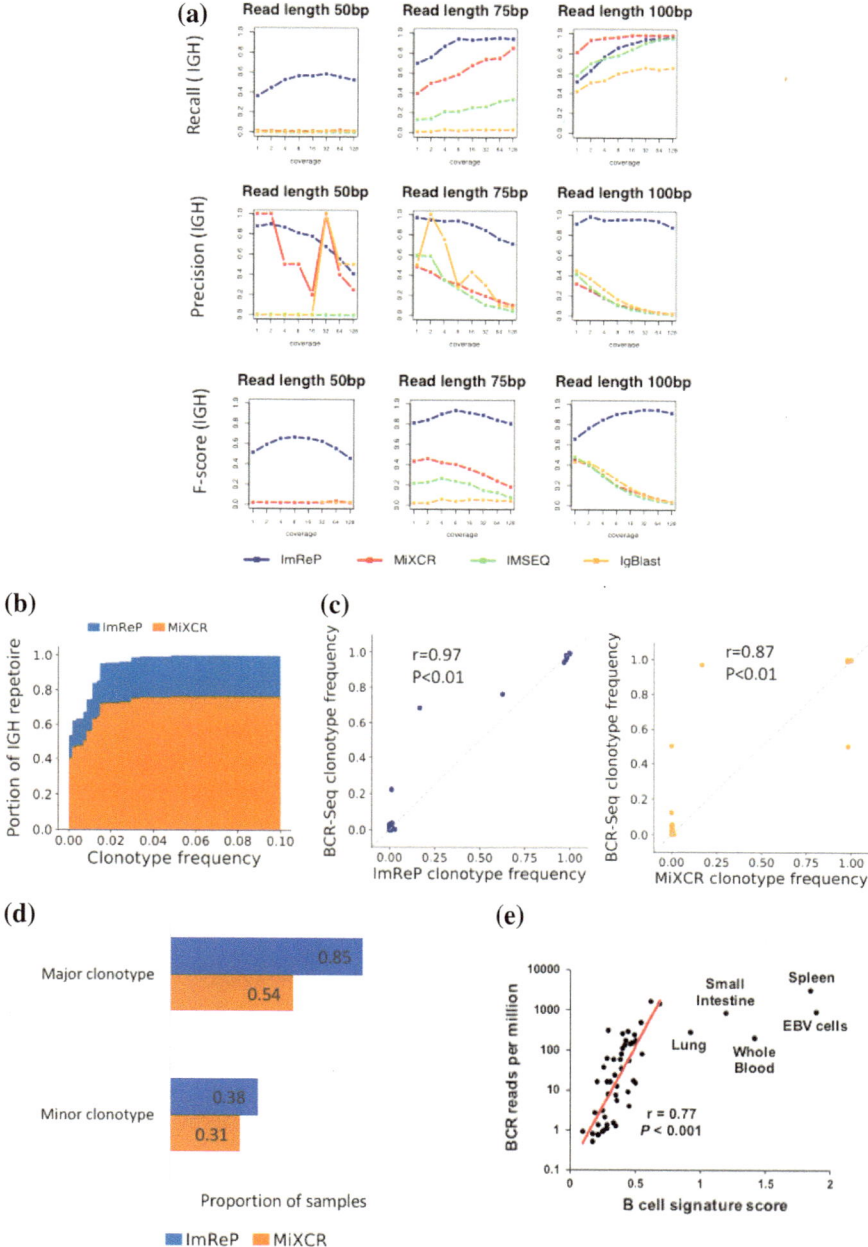

◄**Fig. 38   Evaluation of ImReP. a** Evaluation of ImReP based on the number of assembled CDR3 sequences and comparison to the existing methods. Precision, recall and f-score rates for ImReP (blue), MiXCR (RNA-Seq mode) (red), IMSEQ (green), and IgBlast (orange) on simulated data for immunoglobulin heavy (IGH) transcripts are reported for various reads length (separate plots) and per transcript coverages (1,2,4,8,16,32,64,128) (x-axis). The recall was defined as TP/(TP + FN). Precision is defined as TP/(TP + FP). TP was defined as the number of correctly assembled CDR3 sequence (based on the exact match), FN—the number of true CDR3 sequence not assembled by the method, and FP—the number of incorrectly assembled CDR3 sequences. F-score was defined as the harmonic mean of precision and recall. Ig transcripts were simulated based on the random recombination of V and J gene segments (IMGT database) with non-template insertion at the recombination junction. (b-d) Concordance of targeted BCR-Seq and non-specific RNA-Seq performed on 13 tumor biopsies from Burkitt lymphoma. **b** Area chart shows the proportion of the total IGH repertoire captured by ImRep (blue) and MiXCR (RNA-Seq mode) (orange) depending on the minimum BCR-seq-confirmed clonotypes frequency considered. The *x-axis* corresponds to BCR-seq-confirmed clonotypes frequency Z. The *y*-axis corresponds to the fraction of assembled IGH repertoire with clonotype abundances greater than Z. The total repertoire was defined as the sum of the BCR-seq-confirmed clonotypes abundances. **c** Correlation of IGH clonotype frequencies estimated based on the BCR-Seq data (y-axis) and the RNA-seq data (x-axis) across all the samples. Only clonotypes assembled from RNA-Seq data are presented. **d** ImReP (blue) is able to detect major and minor clonotypes in a larger proportion of the samples compared to MiXCR (RNA-Seq mode) (orange). Major and minor clonotypes were defined based on BCR-Seq data as the clonotype with the largest frequency or smallest frequency, respectively. **e** Correspondence of ImReP-derived reads from Ig receptors to the relative abundance of B cells inferred across 53 GTEx tissues. Scatterplot of the number of all Ig-derived reads per 1 million RNA-Seq reads (y-axis) and B-cell signature score inferred by SaVant based on the gene expression profiles (x-axis)

reads, which were provided for ImReP to assemble IGH clonotypes. Based on the recommendation of the MiXCR, we have provided raw paired-end reads to the tool. BCR-Seq data was generated by Adaptive Biosystems and was analyzed by Adaptive Biosystems's Immune Analyzer package. One difficulty of using BCR-Seq as the gold standard to estimate the efficiency of the RNA-Seq method is that BCR-Seq captures DNA clonotypes, while RNA-Seq only captures the expressed clonotypes. To account for the possible discrepancies, we have first mapped RNA-Seq reads onto the major clonotypes with relative frequency at least 90% detected by BCR-Seq. In five of out 18 BCR-Seq samples, no RNA-Seq reads were mapped to BCR-seq-confirmed major clonotypes. Those samples were excluded from the analysis. In the remaining samples, we have considered the set of CDR3s obtained by BCR-Seq as the total IGH repertoire.

We investigated which portion of the total immune repertoire RNA-Seq is capable of capturing. Using RNA-Seq ImReP was able to capture on average 53.3% of the IGH repertoire, estimated as the sum of the detected BCR-seq-confirmed clonotypes. MiXCR was able to capture 40.1% respectively (Fig. 38b). Overall, ImRep is capable to detect BCR-seq-confirmed clonotypes with a relative frequency exceeding 90% in all of the cases, while MiXCR in 83.3% of cases, respectively. When the frequency of the major clonotype drops below 10%, ImreP was able to detect the major clonotype in 60% of the cases, while MiXCR only in 20% of the cases, respectively. Remarkably both methods were able to detect major clonotype with the frequency below 1% in one of the samples (Table 14). We have also investigated the ability of both methods

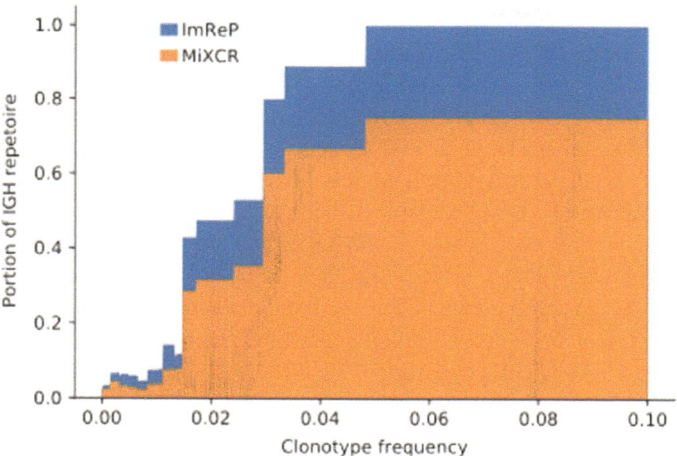

Fig. 39 Concordance of targeted BCR-Seq and non-specific RNA-Seq performed on 13 tumor biopsies from Burkitt lymphoma. Area chart shows the proportion of the total IGH repertoire captured by ImRep (blue) and MiXCR (RNA-Seq mode) (orange) depending on the minimum BCR-seq-confirmed clonotypes frequency considered. The *x-axis* corresponds to BCR-seq-confirmed clonotypes frequency Z. The *y*-axis corresponds to the fraction of assembled IGH repertoire with clonotype abundances greater than Z. The total repertoire was defined as the total number of the BCR-seq-confirmed clonotypes

to detect BCR-seq-confirmed minor clonotypes. The average frequency of the minor clonotypes across all samples was 0.37%. ImreP was able to detect minor clonotype in 38% of the samples (Fig. 38d). Despite the ability of ImReP and MiXCR to capture the majority of BCR-seq-confirmed repertoire, both methods often missed the rare clonotypes due to the limited number of BCR-derived reads in RNA-Seq data. ImReP was able to detect 50% of all BCR-Seq-confirmed clonotypes with the relative frequency higher than 0.24%. MiXCR was able to detect 50% of all BCR-Seq-confirmed clonotypes with the relative frequency higher than 0.29% (Fig. 39). Both methods were able to accurately estimate the relative frequencies of assembled clonotypes (ImRep : $r = 0.97$, *p*-value $= 10^{-40}$; MiXCR $r = 0.87$, *p*-value $= 10^{-15}$) (Fig. 38c).

We further validated the ability of ImReP to accurately infer the proportion of immune cells in the sampled tissue. We hypothesized that the fraction of B cells in the sample will be proportional with the fraction of receptor-derived reads in our RNA-Seq data. We used a transcriptome-based computational method, SaVant (Lopez et al., 2017), which uses cell-specific gene signatures (independent of *Ig* transcripts) to infer the relative abundance of B cells within each tissue sample (Table 16). The B cell signature used by SaVant are derived from CD19+ cells and might not represent every B cell subset (Landsverk et al., 2017). However, CD19+ cells likely represent the largest populations of B cell subsets and many of the CD19 negative B cell subsets may still have a similar gene signature to the CD19 signatures. We found that B cell signatures inferred by SaVant showed the positive correlation with the size of IGH repertoire ($r = 0.77$, $P < 0.001$) (Fig. 38e). An exception to this correlation was for tissues that contain the highest density of B cells: spleen, whole blood, small intestine (terminal ileum), lung, and EBV-transformed lymphocytes (LCLs).

ImReP identified over 26 million reads overlapping 3.6 million distinct CDR3 sequences that originate from diverse human tissues. The majority of assembled CDR3 sequences derived from immunoglobulin heavy chain (IGH) (1.7 million), 0.9 million were derived from the immunoglobulin kappa chain (IGK), and 1.0 million from the immunoglobulin lambda chain (IGL). The vast majority of all assembled CDR3s had a low frequency in the data. 98% of CDR3 sequences had a count of less than 10 reads, and the median CDR3 sequence count was 1.4. CDR3 sequences derived from IGK were the most abundant across all tissues, accounting on average for 54% of the entire B-cell population (Fig. 40).

We compared the length and amino acid composition (Crooks, Hon, Chandonia, & Brenner, 2004) of the assembled CDR3 sequences of *Ig* receptor chains (Fig. 41a). Consistent with previous studies, we observed that immunoglobulin light chains have notably shorter and less variable CDR3 lengths compared to heavy chains (Philibert et al., 2007). The tissue type appears to have no effect on the length distribution of CDR3 sequences (Fig. 42). In line with other studies (Philibert et al., 2007; Hoi & Ippolito, 2013), both light chains exhibited a reduced amount of sequencing diversity (Fig. 41b).

We observed per sample an average of 1331 distinct Ig clonotypes. We normalized the number of distinct clonotypes by the total number of raw RNA-Seq reads, which we call number of clonotypes per one million raw RNA-Seq reads (CPM). As the number of distinct clonotypes does not increase linearly with the sequencing depth, CPM metric should not be used in studies comparing clonotype diversity across various phenotypes. Instead, CPM is intended to be an informative measure of clonal diversity adjusted for sequencing depth. One technique allowing to properly adjust for sequencing depth per sample is to subsample reads, to have identical number of reads per sample.

We used per sample alpha diversity (Shannon entropy) to incorporate the total number of distinct clonotypes and their relative frequencies into a single diversity metric. Among all tissues, spleen has the largest B-cell population, with a median of 1301 Ig-derived reads per one million RNA-Seq reads. It also has the most diverse population of B cells with median per sample alpha diversity of 7.6 corresponding to 1025 CPM (Fig. 43; Table 15). Organs that possess mucosal, exocrine, and endocrine sites (n = 24) harbor a rich clonotype population with a median of 87 CPM per sample. Minor salivary glands have the highest IG diversity in the group (alpha = 7.1) and surpass the diversity of the terminal Ileum containing Peyer's Patches, which are secondary lymphoid organs (Table 15).

Tissues not related to the immune system, including adipose, muscle, and the organs from the central nervous system, contained a median of 6 CPM per sample, which are most likely due to the blood content of the tissues (Yu et al., 1991). The highest number of distinct CDR3 sequences among non-lymphoid organs was present in the omentum, a membranous double layer of adipose tissue containing fat-

Table 16 The adjusted clonotypic richness of B cells, calculated as the number of distinct amino acid sequences of CDR3 per one million RNA-Seq reads (CPM) normalized by the proportion of the B cell in the sample

Tissue type	Adjusted CPM (IGH)	Adjusted CPM (IGK)	Adjusted CPM (IGL)
Adipose—Subcutaneous	0.31	0.10	3.33
Adipose—Visceral (omentum)	1.22	0.18	2.92
Adrenal Gland	0.34	0.13	3.76
Artery—Aorta	1.16	0.21	4.54
Artery—Coronary	1.19	0.21	3.86
Artery—Tibial	0.26	0.13	8.41
Bladder	2.47	0.23	2.86
Brain—Amygdala	0.10	0.08	6.78
Brain—Anterior cingulate cortex (BA24)	0.14	0.07	7.30
Brain—Caudate (basal ganglia)	0.13	0.07	6.58
Brain—Cerebellar hemisphere	0.15	0.10	12.45
Brain—Cerebellum	0.18	0.12	13.11
Brain—Cortex	0.14	0.07	7.87
Brain—Frontal cortex (BA9)	0.13	0.08	7.89
Brain—Hippocampus	0.11	0.09	5.99
Brain—Hypothalamus	0.10	0.07	6.27
Brain—Nucleus accumbens (basal ganglia)	0.12	0.07	6.86
Brain—Putamen (basal ganglia)	0.13	0.08	6.85
Brain—Spinal cord (cervical c-1)	0.10	0.07	4.60
Brain—Substantia nigra	0.09	0.07	5.49
Breast—Mammary tissue	3.91	0.37	3.40
Cells—EBV-transformed lymphocytes	0.22	0.04	1.12
Cells—Transformed fibroblasts	0.10	0.16	12.75
Cervix—Ectocervix	1.75	0.34	3.18
Cervix—Endocervix	2.12	0.39	2.32
Colon—Sigmoid	0.78	0.17	5.93
Colon—Transverse	12.37	0.74	3.21
Esophagus—Gastroesophageal junction	0.16	0.11	5.62

(continued)

Table 16 (continued)

Tissue type	Adjusted CPM (IGH)	Adjusted CPM (IGK)	Adjusted CPM (IGL)
Esophagus—Mucosa	2.45	0.39	6.18
Esophagus—Muscularis	0.15	0.11	5.38
Fallopian tube	2.79	0.33	1.86
Heart—Atrial appendage	0.26	0.13	5.78
Heart—Left ventricle	0.20	0.13	6.53
Kidney—Cortex	1.00	0.20	3.47
Liver	1.23	0.32	5.24
Lung	3.87	0.32	2.27
Minor Salivary gland	22.36	1.18	4.37
Muscle—Skeletal	0.22	0.15	11.09
Nerve—Tibial	0.27	0.12	4.04
Ovary	0.54	0.21	6.73
Pancreas	0.59	0.25	7.75
Pituitary	0.27	0.13	4.44
Prostate	0.86	0.22	3.60
Skin—Not sun exposed (Suprapubic)	0.37	0.19	6.39
Skin—Sun exposed (lower leg)	0.20	0.15	6.87
Small Intestine—Terminal ileum	8.35	0.53	1.94
Spleen	10.10	0.53	1.29
Stomach	5.95	0.50	2.93
Testis	0.13	0.08	5.45
Thyroid	1.37	0.19	4.43
Uterus	0.52	0.23	5.33

We have used SaVant, a transcriptome-based computational method to infer the relative abundance of B cells within each tissue sample based on cell-specific gene signatures (independent of *Ig* transcripts)

associated lymphoid clusters. As expected (De Rossi et al., 1990), Epstein–Barr virus (EBV)-transformed lymphocytes (LCL) harbored a large homogeneous population of *Ig* clonotypes (Table 15; Fig. 44). The number of reported clonotypes was normalized by the proportion of B cells within each tissue sample (Table 16). We have used SaVant to infer the relative abundance of B cells within each tissue sample based on cell-specific gene signatures (independent of *Ig* transcripts).

Amino acid sequences of clonotypes exhibited extreme inter-individual dissimi-larity, with 88% of clonotypes unique to a single individual (private) (Fig. 45a). The remaining ~400,000 clonotypes were shared by at least two individuals (public). A small fraction of B cells in many tissues limits our ability to capture the entire Ig repertoire and thus resulting in classifying some public clonotypes as private. The number of individuals sharing clonotypes varied across Ig chains, with immunoglobulin light chains having the highest number of public clonotypes. Twenty-five percent of all IGK clonotypes were public, and the number of individuals sharing the IGK clonotype sequences can be as high as 471 (Fig. 45b). The limited capacity of RNA-Seq to cover low abundant clonotypes may misclassify public clonotypes as private. Consistent with the previous studies (Li et al., 2016; Warren et al., 2011), we observe public clonotypes to be significantly shorter in length than the private ones (p-value $< 2 \times 10^{-16}$). For example, IGH chain public clonotypes had an average length of 13 amino acids, and private clonotypes had an average length of 16. We also examined whether the public clonotypes were more often shared across tissues within an individual. Only 14% of the ~ 240,000 clonotypes shared across tissues were public. The majority of clonotypes were individual- and tissue-specific (Fig. 45c).

A large number of individuals available through the study allow us to establish a pairwise relationship between the tissues and track the flow of Ig clonotypes across human tissues. We observed a significant increase in the number of CDR3 sequences shared across pairs of tissues from the same individuals. Further, we observed this pattern consistently for all chains of Ig receptors (p-value $< 2 \times 10^{-16}$) (Fig. 46a; Table 17). We observe a different amount of shared CDR3 sequences across different types of Ig chains with an increase in immunoglobulin light chains compared to *Ig* heavy chain. On average, we observe 7.0 CDR3 sequences to be shared across a pair of tissues from the same individuals. Pairs of tissues from different individuals share on average 3.6 CDR3 sequences (Fig. 46a; Table 17).

To establish the flow of *Ig* clonotypes across various tissues, we compared clonotype populations between and within the same individuals. We limited this analysis to pairs of tissues for which we had at least 10 individuals (870 pairs of tissues out of 1378 possible pairs). We used beta diversity (Sørensen–Dice similarity index) to measure compositional similarities between the tissues in terms of gain or loss of CDR3 sequences (Fig. 46b). For the majority of the 870 available tissue pairs, we observe no *IGH* sequences in common, which corresponds to beta diversity of 0.0.

We have examined the flow of *IGH* clonotypes across tissues and presented it as a network (Fig. 46b). Among 870 available tissue pairs, we have identified 56 tissue pairs with beta diversity above 0.001. The spleen was the most highly connected tissue, with 17 connections, followed by lung, with 16 connections. Clonotypes represents one connected component, meaning that every two nodes are connected directly or via other nodes. Clonotype populations of spleen and lung are the most similar (0.02 beta diversity), other pairs include minor salivary gland and esophagus mucosa, terminal ileum (small intestine) and transverse colon. We observe above 200 pairs of tissues with beta diversity above 0.001 for immunoglobulin light chains (Figs. 47 and 48). The most similar tissue pairs for *IGK* chain were spleen and transverse colon (0.15 beta diversity).

Histological images of tissue cross-sections and pathologists' notes have been used to validate the ImReP's ability to detect the samples with a high lymphocyte

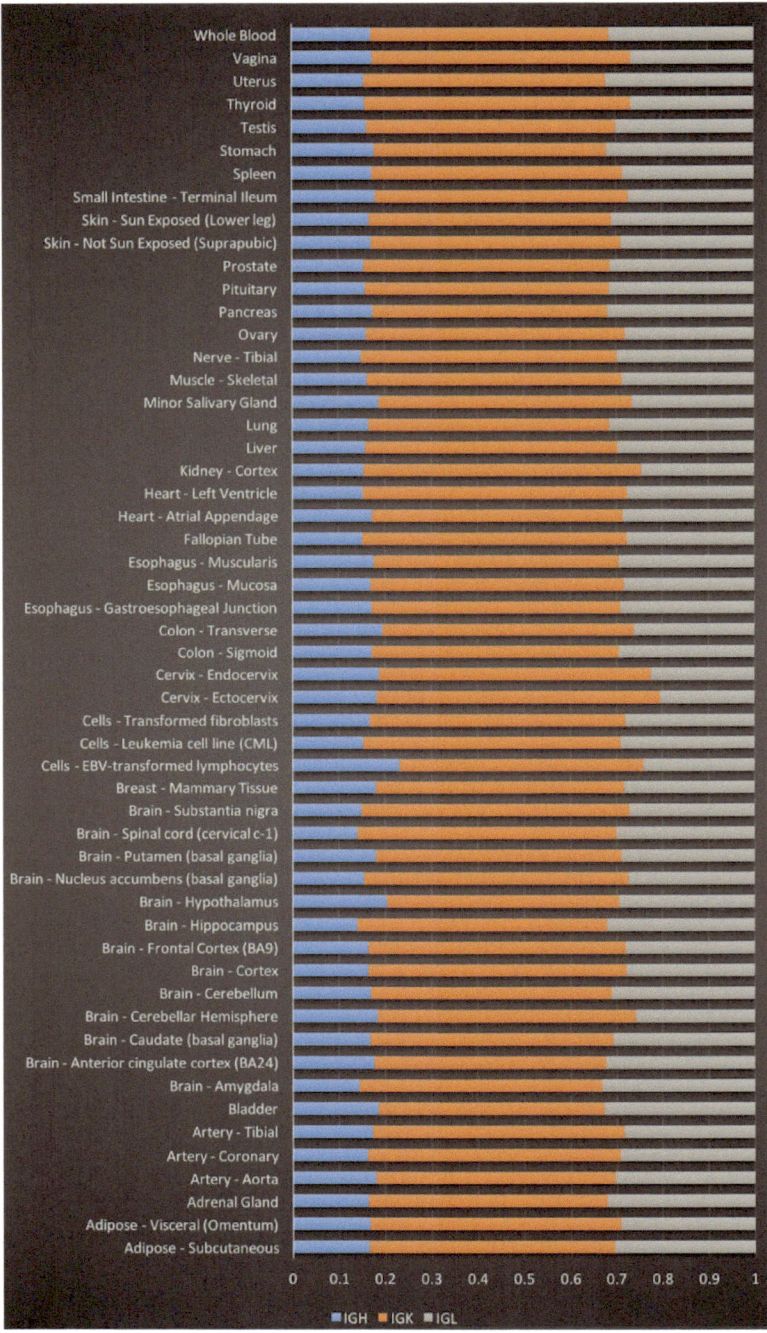

Fig. 40 The fraction of IGH, IGK, and IGL among the whole B cell population across 53 body sites. The fraction is calculated based on the number of reads derived from CDR3 sequences of each BCR chain

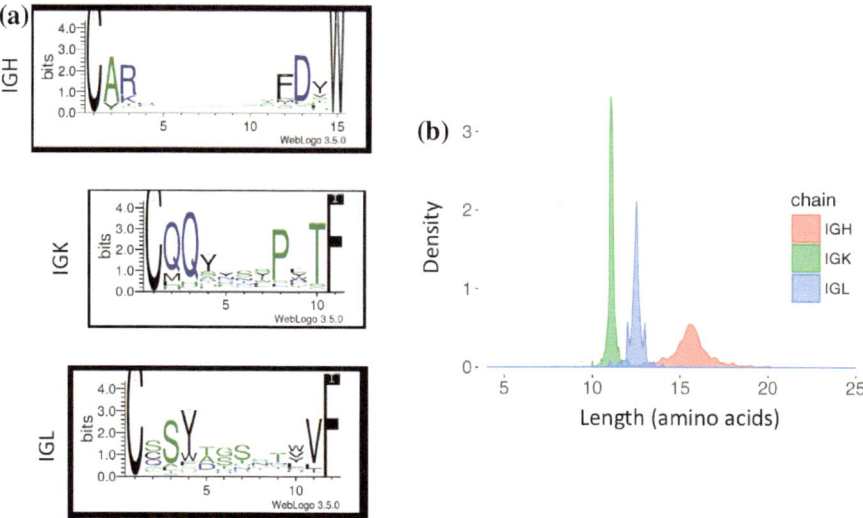

Fig. 41 The length and amino acid composition of the assembled CDR3 sequences of immunoglobulin receptor chains. The sequence logo (using WebLogo) of amino acid composition representation for CDR3 sequences of mean length. The height of the amino acid within the stack indicates the relative frequency. Distribution of CDR3 sequence length is estimated using kernel density. **a** Sequence logo of a 15-amino-acid CDR3 sequence of IGH. **b** Sequence logo of 11-amino-acid CDR3 of IGK. **c** Sequence logo of a 12-amino-acid CDR3 sequence of IGL. Distribution of CDR3 sequence length is estimated using s kernel density separately for each Ig chain

content, which often corresponds to a disease state. We examined the IGH clonotype populations from thyroid tissue across individuals. The median number of inferred distinct CDR3 sequences per sample was 20, though 14.5% of the samples had more than 500 distinct CDR3 sequences. We observed the highest number of CDR3 sequences among all the thyroid samples in an individual with late-stage Hashimoto's thyroiditis, an autoimmune disease characterized by lymphocyte infiltration and T-cell mediated cytotoxicity. According to pathologists' notes, Hashimoto's disease was present in 11.2% of thyroid samples, with varying degrees of severity. First, we used pathologists' notes to annotate samples as healthy (n = 183) or bearing Hashimoto's disease (n = 23), and then we compared the adaptive repertoire diversity between these groups. We observed a significant increase in the number of distinct IGH clonotypes in samples with Hashimoto's thyroiditis (p-value = 1.5×10^{-5}) (Fig. 49). The number of clonotypes varied from 113 for focal Hashimoto's thyroiditis to 5621 for late-stage Hashimoto's thyroiditis (Fig. 50a). In addition, high clonotype diversity in kidney samples indicated the presence of glomerulosclerosis. In lung samples, high clonotype diversity corresponded to inflammatory diseases such as sarcoidosis and bronchopneumonia.

We observed no difference in clonal diversity in males and females across the tissues, except in breast tissues (p-value < 3.2×10^{-12}). Increased clonotype diver-

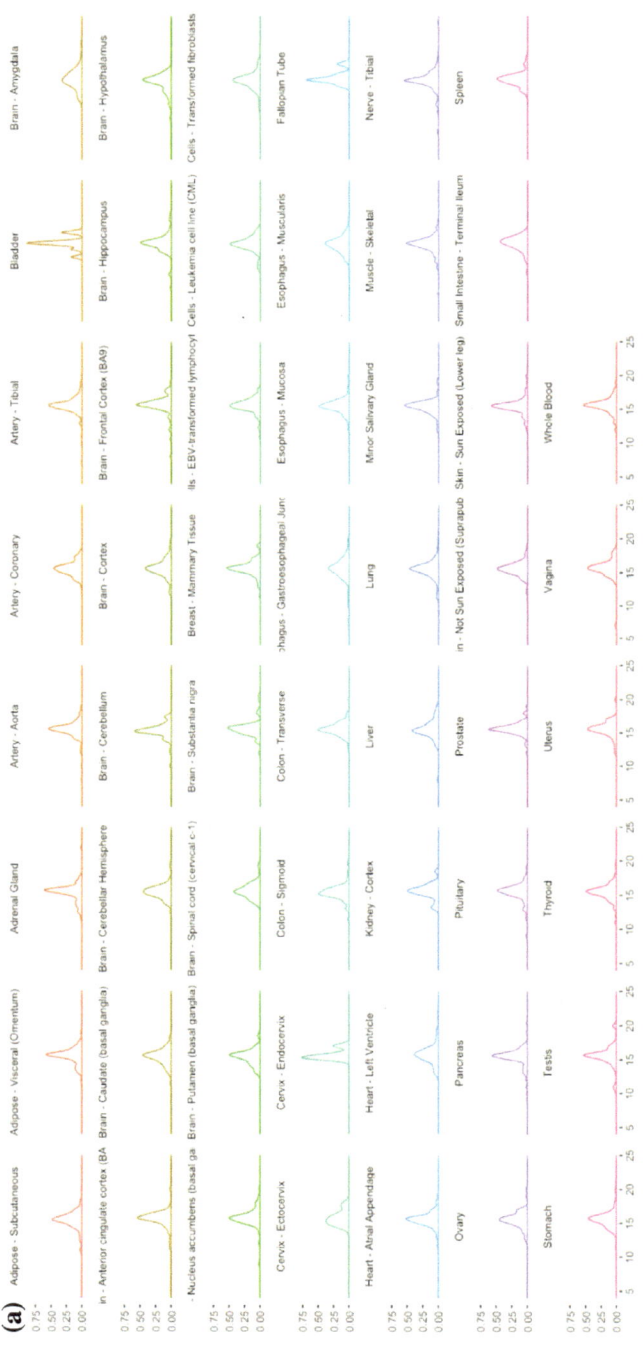

Fig. 42 a Length distribution of amino acid sequences of the CDR3 region in immunoglobulin heavy chain (IGH), presented across 53 various body sites. **b** Length distribution of amino acid sequences of the CDR3 region in immunoglobulin kappa chain (IGK), presented across 53 various body sites. **c** Length distribution of amino acid sequences of the CDR3 region in immunoglobulin kappa chain (IGK), presented across 53 various body sites

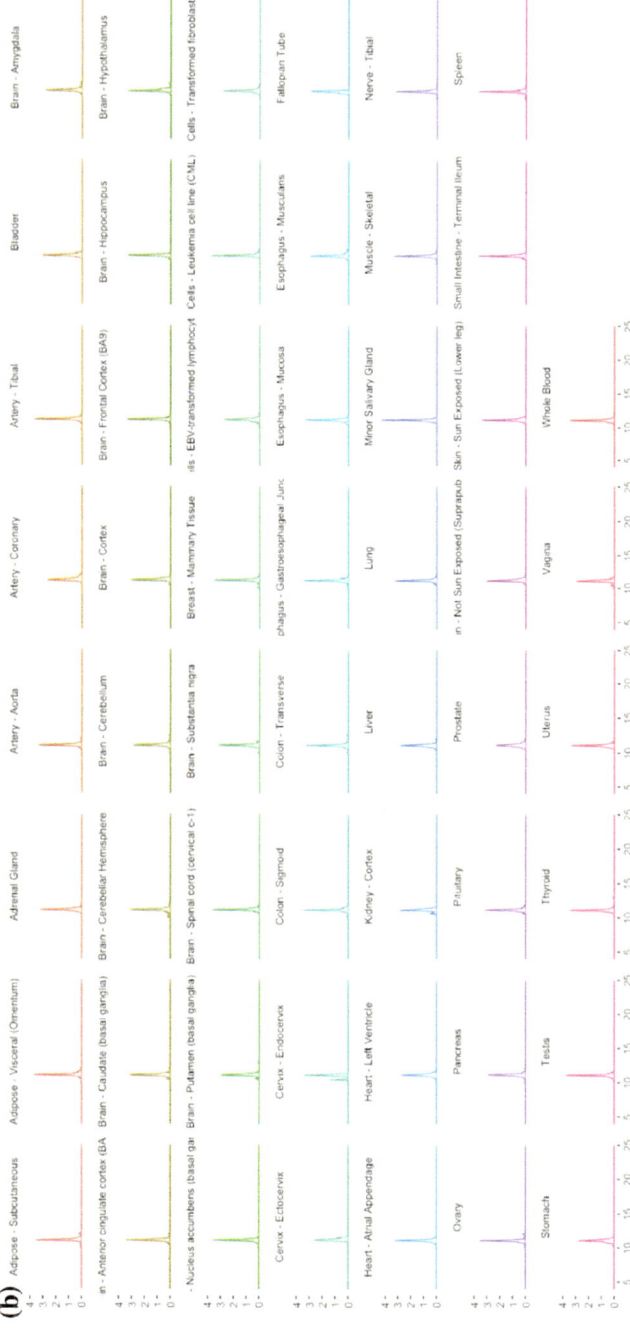

Fig. 42 (continued)

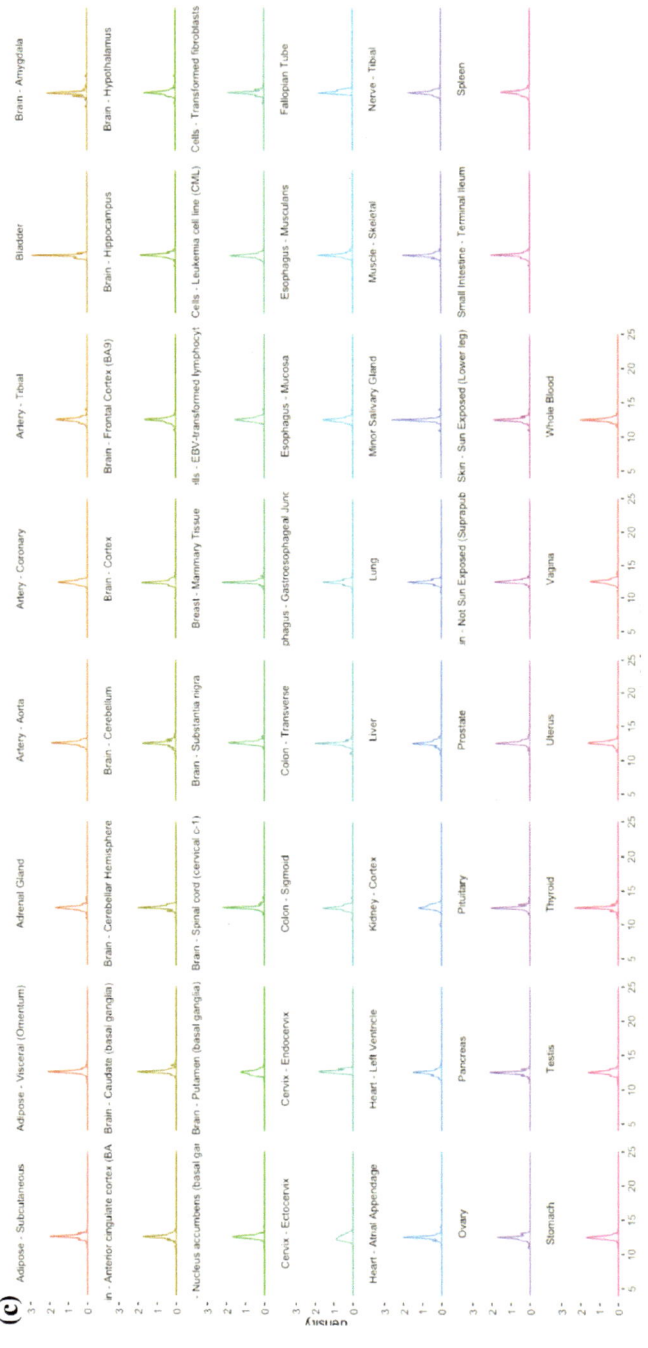

Fig. 42 (continued)

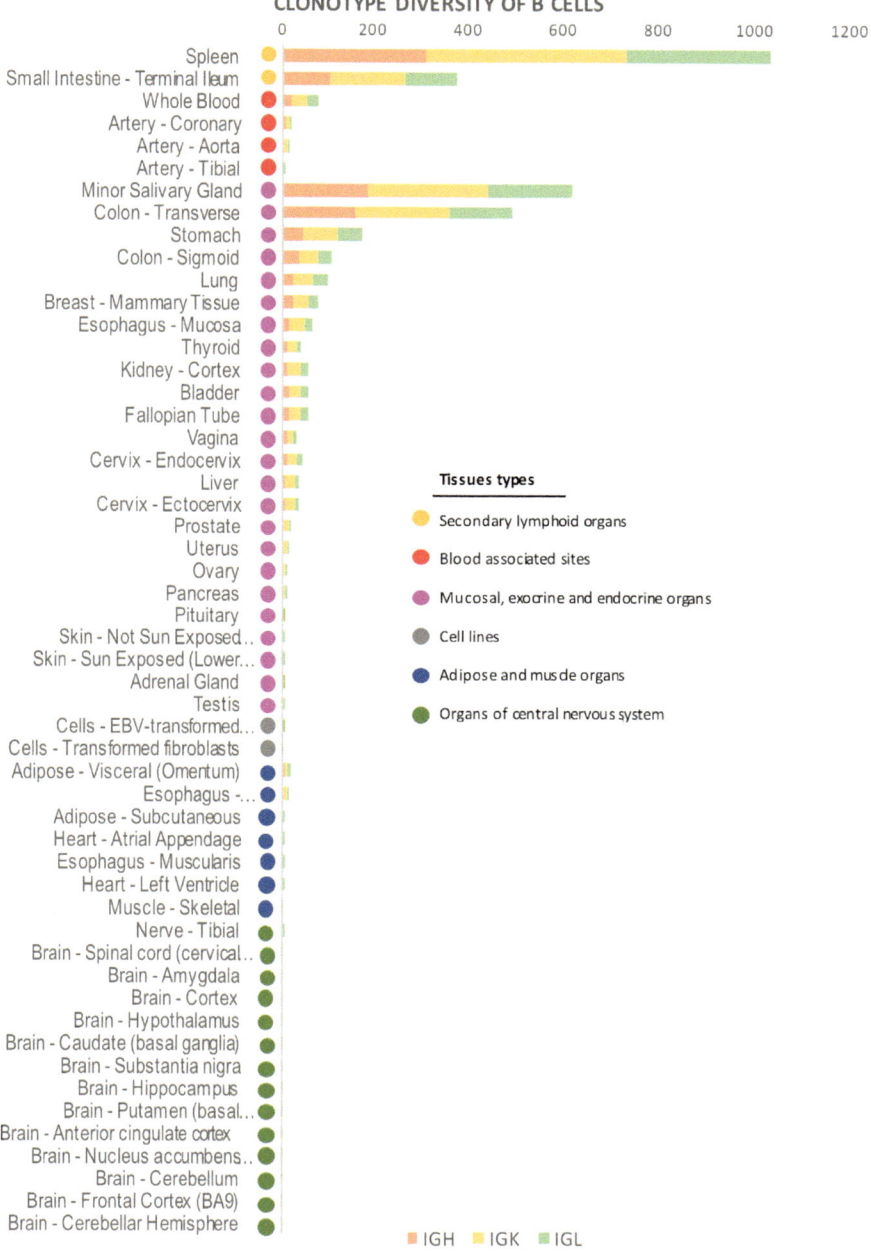

CLONOTYPE DIVERSITY OF B CELLS

◄**Fig. 43** Adaptive immune repertoires across multiple human tissues. Adaptive immune repertoires of 8555 samples across 544 individuals from 53 body sites obtained from Genotype-Tissue Expression study (GTEx v6). We group the tissues by their relationship to the immune system. The first group includes the lymphoid tissues (n = 2, red colors). The second group includes blood associated sites including whole blood and blood vessel (n = 4, red color). The third group are the organs that encompass mucosal, exocrine and endocrine organs (n = 21, lavender color). The fourth group are cell lines (n = 3, grey color). The fifth group are adipose or muscle tissues and the gastroesophageal junction (n = 7, blue color). The sixth group are organs from the central nervous system (n = 14, green color). Histogram reports clonotypic richness of B cells, calculated as the number of distinct amino acid sequences of CDR3 per one million RNA-Seq reads (CPM). The median number of distinct amino acid sequences of CDR3 are presented individually for immunoglobulin heavy chain (IGH), immunoglobulin kappa chain (IGK), immunoglobulin lambda chain (IGL)

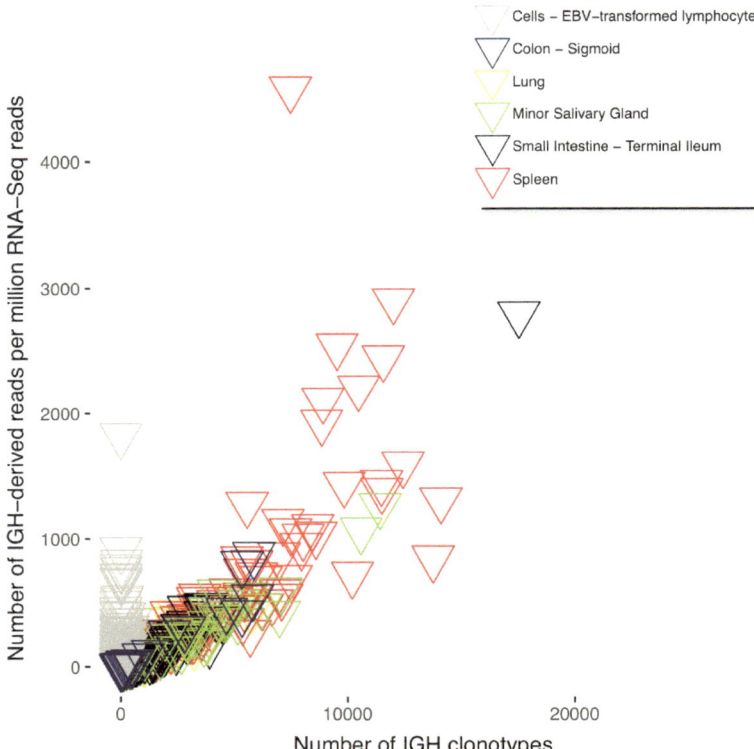

Fig. 44 Scatterplot of the number of IGH clonotypes (CDR3s) in each sample, plotted against the number of IGH-derived reads per 1 million RNA-Seq reads

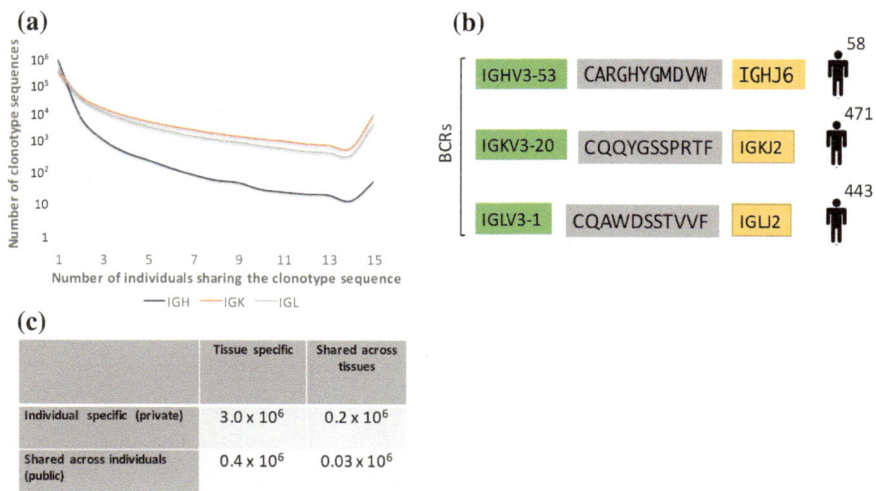

Fig. 45 Private and public Ig clonotypes. **a** Distribution of frequencies of private (n = 1) and public (n > 1) clonotypes across 544 individuals. We collect clonotypes from all tissues of the same individual into a single set corresponding to that individual. **b** The most public clonotypes (shared across the maximum number of individuals) and corresponding VJ recombination are presented for IGH, IGK, IGL. **c** Clonotypes sequences are classified into public clonotypes (shared across individuals), private (individual-specific), tissue-specific, and clonotypes shared across multiple tissues. The number of clonotypes falling into each pair of categories is reported across Ig receptor chains

sity of breast tissue in male individuals corresponded to gynecomastia, a common disorder of non-cancerous enlargement of male breast tissue (Fig. 50b).

Our study is the first that systematically accounts for almost all reads, totaling one trillion, available via three RNA-Seq datasets. We demonstrate the value of analyzing unmapped reads present in the RNA-Seq data to study the non-co-linear, RNA editing, immunological, and microbiome profiles of a tissue. We developed a new tool (ROP) that accounts for 99.9% of the reads, a substantial increase compared over the 82.2% of reads account for using conventional protocols. We found that the majority of *unmapped reads* are human in origin and from diverse sources, including repetitive elements, A-to-I RNA editing, circular RNAs, gene fusions, trans-splicing, and recombined B and T cell receptor sequences. In addition to those derived from human RNA, many reads were microbial in origin and often occurred in numbers sufficiently large to study the taxonomic composition of microbial communities in the tissue type represented by the sample.

We found that both unmapped human reads and reads with microbial origins are useful for differentiating between type of tissue and status of disease. For example, we found that the immune profiles of asthmatic individuals have decreased immune diversity when compared to those of controls. Further, we used our method to show that immune diversity is inversely correlated with microbial load. This case study highlights the potential for producing novel discoveries, when the information in

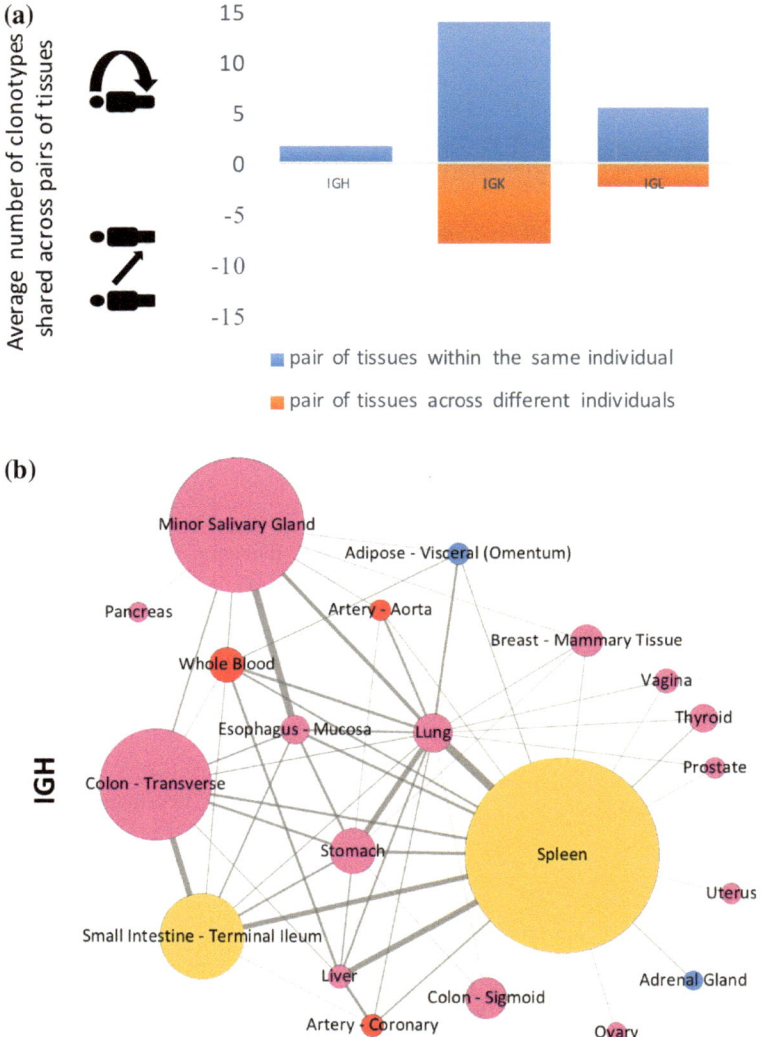

Fig. 46 The flow of *Ig* clonotypes across diverse human tissues. Results are based on pairs of tissues with at least 10 individuals. **a** The number of *Ig* clonotype sequences shared across pairs of tissues from the same individuals (blue color) and from different individuals (orange color) is presented. **b–c** Flow of clonotypes across diverse human tissues is presented as a network. Each node is a tissue with the size proportional to a median number of clonotypes of the tissue. The color of the node corresponds to a type of the tissue type: lymphoid tissues (yellow colors), blood associated sites (red color), organs that encompass mucosal, exocrine and endocrine organs (lavender color). Compositional similarities between the tissues in terms of gain or loss of CDR3 sequences are measured across valid pairs of tissues using beta diversity (Sørensen–Dice similarity index). Edges are weighted according to the beta diversity. **b** The flow of *IGH* clonotypes across diverse human tissues is presented as a network. Edges with beta diversity > 0.001 are presented

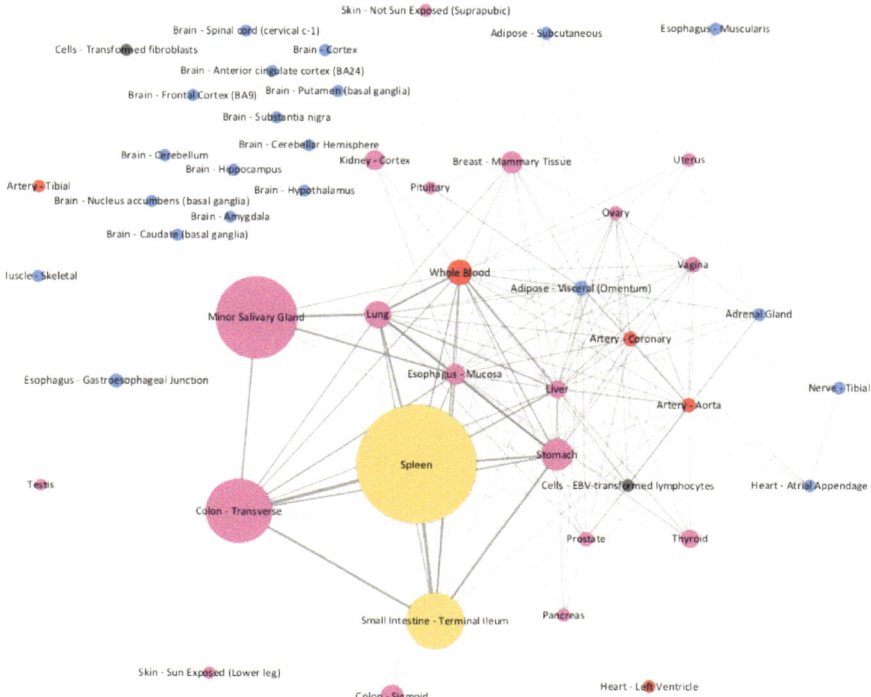

Fig. 47 The flow of IGK clonotypes across diverse human tissues is presented as a network. Edges with beta diversity > 0.001 are presented

Table 17 Data used for Fig. 45a represented as a table

	Number of CDR3 sequences shared across pars of tissues	
Chain type	The same individuals	Different individuals
IGH	1.5	0.01
IGK	13.9	8.2
IGL	5.4	2.4

Results are based on pairs of tissues with at least 10 individuals. The number of clonotype sequences shared across pairs of tissues from the same individuals is presented in column 2. The number of clonotype sequences shared across pairs of tissues from different individuals is presented in column 3

RNA-Seq data is fully leveraged by incorporating the analysis of unmapped reads, without need for additional TCR/BCR or microbiome sequencing. The ROP profile of unmapped reads output by our method is not limited to RNA-Seq technology and may apply to whole-exome and whole-genome sequencing. We anticipate that ROP

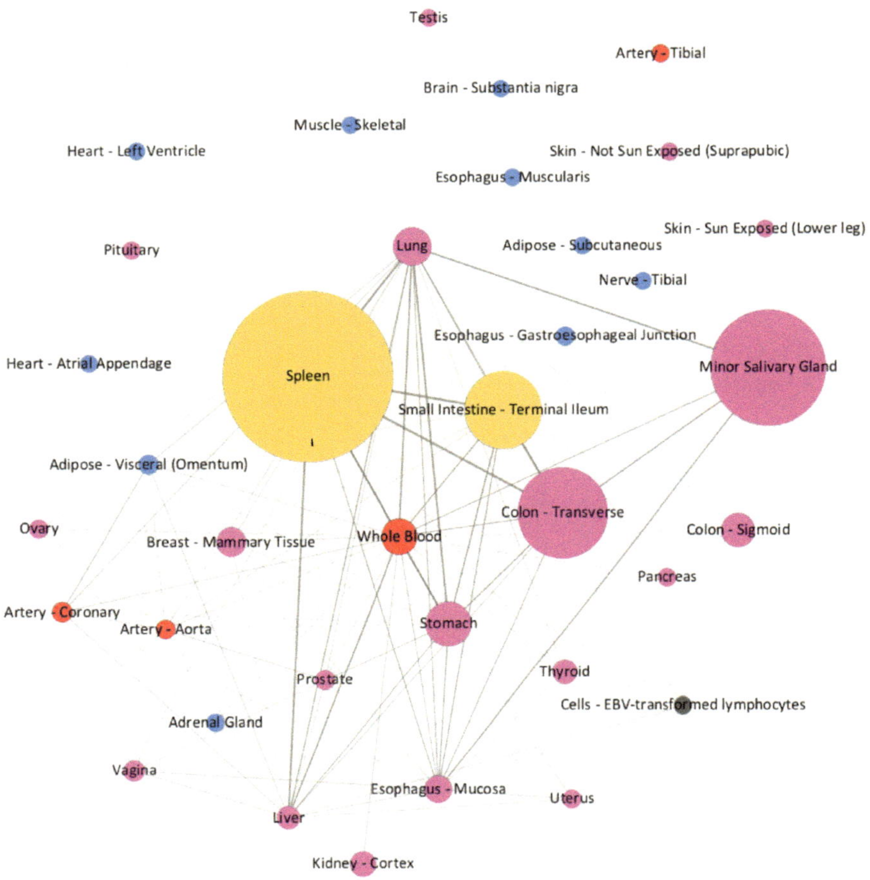

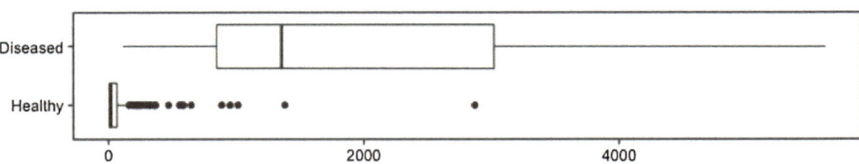

Fig. 48 The flow of IGL clonotypes across diverse human tissues is presented as a network. Edges with beta diversity > 0.001 are presented

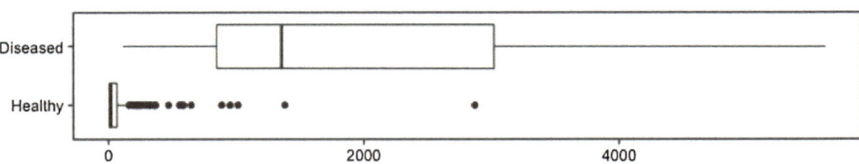

Fig. 49 The number of IGH clonotypes for healthy individuals (Healthy) and individuals bearing Hashimoto's disease (diseased). Pathologists' notes were used to annotate samples as healthy or diseased. A significant increase in the number of distinct IGH clonotypes in samples with Hashimoto's thyroiditis (p-value $= 1.5 \times 10^{-5}$) is observed

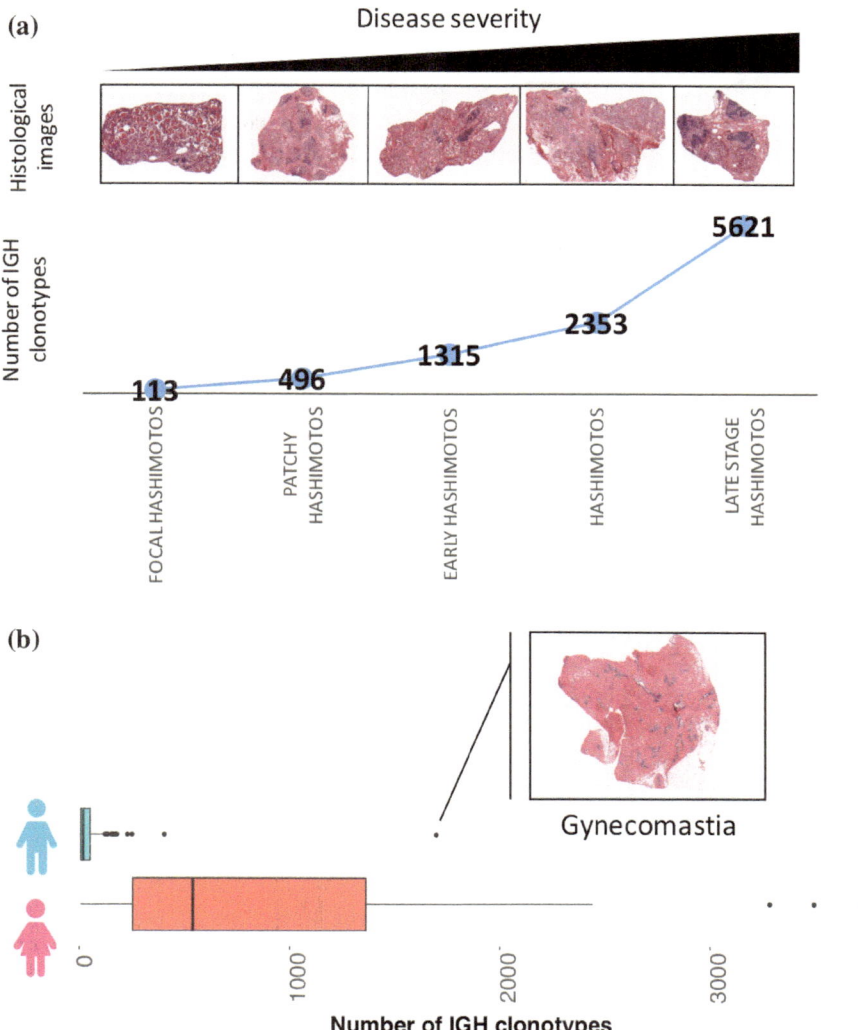

Fig. 50 ImReP is able to identify samples with high activity of lymphocytes. Histological images of tissue cross-sections and pathologists' notes have been used to validate the ImReP's ability to detect the samples with a high activity of lymphocytes. **a** Samples were ordered by Hashimoto's thyroiditis severity, as reported by pathologists' notes. Histological images are provided to illustrate the disease state. The average number of Ig clonotypes is reported for each disease group. **b** Boxplot reporting number of clonotypes in the breast tissues for males and females. The outlier among the male samples is illustrated with the histological image

profiling will have broad future applications in studies involving different tissue and disease types.

We observed large effects when using different library preparation protocols on non-co-linear, immunological, and microbial profiles. For example, the poly(A) protocol better captures antibody transcripts by enriching for polyadenylated transcripts, while ribo-depletion protocols capture more circRNAs. The results presented here suggest that selection of a protocol impacts quality of analysis results, and our study may guide the choice of protocol depending on the features of interest.

The ROP protocol facilitates a simultaneous study of immune systems and microbial communities, and this novel method advances our understanding of the functional, interrelated mechanisms driving the immune system, microbiome, human gene expression, and disease etiology. Moreover, we demonstrate the in-depth analysis of microbiome and immune reads can elucidate important biological and clinical information from RNA-Seq data. Firstly, we used high throughput RNA sequencing from whole blood to perform microbiome profiling and identified an increased diversity in schizophrenia patients.

While other studies of human microbiome using RNA-Seq have been conducted (Croucher & Thompson, 2010; McClure et al., 2013), this is the first assessing the microbiome from whole blood by using unmapped non-human reads. Despite the fact that transcripts are present at much lower fractions than human reads, we were able to detect microbial transcripts from bacteria and archaea in almost all samples. The microbes found in blood are thought to be originating from the gut as well as oral cavities (Potgieter, Bester, Kell, & Pretorius, 2015; Spadoni et al., 2015), which is in line with our finding that the microbial profiles found in our study most closely resemble the gut and oral microbiome as profiled by the HMP (Human MIcrobiome Project, 2012). The taxonomic profile of the cohort samples suggests the prevalence of the several phyla, Proteobacteria, Firmicutes and Cyanobacteria, across individuals and different disorders included in our study. This is in line with a recent study that used 16S targeted meta-genomic sequencing, which reported Proteobacteria and Firmicutes among the most abundant phyla detected in blood (Païssé et al. 2016).

Rigorous quality control is critically important for any high-throughput sequencing project, especially in the context of studying the microbiome (Salter et al., 2014). To this end, we performed both negative and positive quality control experiments, and we carefully evaluated possible contamination effects introduced during the experiments. Our results suggest that the detected phyla represent true microbial communities in whole blood and are not due to contaminants. However, it should be noted that whether only the microbial products crossed into the bloodstream or whether the microbes themselves are present in blood cannot be answered using sequencing techniques. Future experiments, for example, using microscopy, culturing or direct measures of gut permeability, may be able to shed light on this question.

The most striking finding of our study that relates to diseases affecting the central nervous system is the increased microbial alpha diversity in schizophrenia patients compared to controls and the other two disease groups (ALS, BPD). We replicate this finding in an independent cohort of schizophrenia cases and controls. The replication experiment, while based on different library preparation (Ribo-Zero versus

Poly(A)), provides strong evidence for an increased alpha diversity of the microbiome detected in blood in schizophrenia and explains roughly 5% of disease variation. We not only observe an increased individual microbial diversity, but also an increased diversity between individuals (Beta diversity) with schizophrenia compared to controls, rendering it unlikely that a single phylum or microbial profile is causing the disease-specific increase in diversity. Nevertheless, in our study we observed that two phyla in particular, Planctomycetes and Thermotogae, were present in significantly more schizophrenia samples when compared to the other groups. Interestingly, Planctomycetes is group of gram-negative bacteria closely related to Verrucomicrobia and Chlamydiae; together these comprise the Planctomycetes-Verrucomicrobia-Chlamydiae superphylum (Hou et al., 2008). From peripheral blood, infection with Chlamydiaceae species has been reported to be increased in schizophrenia (40%) compared to controls (7%) (Fellerhoff, Laumbacher, Mueller, Gu, & Wank, 2006). Since Chlamydiae is one of the taxa of the superphylum, it is possible that the increase in Planctomycetes we observe is related to the observed increase in Chlamydiaceae species. As the collection of available reference genomes continues to grow and improve, future studies are needed to corroborate and refine these findings.

For the study of microbiome diversity, we employed reference-based methods (PhyloSift and MethPhlAn) and the EMDebruin method, a purely reference-agnostic approach. The latter showed strong correspondence to both reference-based methods, highlighting the value of this unbiased sequence-based analysis for investigating microbial differences across groups. However, in addition to differences in distribution of microbial transcripts, EMDebruin may capture variation of other, yet unknown, origin.

In addition to our observation that microbial diversity is more generally increased in schizophrenia, our study demonstrates the value of analyzing non-human reads present in the RNA-Seq data to study the microbial composition of a tissue of interest (Jorth et al., 2014, Kostic et al., 2011). The RNA-Seq approach avoids biases introduced by primers in targeted 16S ribosomal RNA gene profiling. In addition, since *mRNA stability is low in prokaryotes*, RNA-Seq might offer a potential advantage of avoiding contamination of genomic DNA by dead cells compared to genome sequencing (Ben-Amor et al., 2005). Given the many large-scale RNA-Seq datasets that are becoming available, we anticipate that high-throughput meta-transcriptome-based microbiome profiling will find broad applications as a hypothesis-generating tool in studies across different tissues and disease types.

The increased microbial diversity observed in schizophrenia could be part of the disease etiology (i.e., causing schizophrenia) or may be a secondary effect of disease status. In our sample, we observed no correlation between increased microbial diversity and genetic risk for schizophrenia as measured by polygenic risk scores (Ripke et al., 2013). In addition, it is remarkable that bipolar disorder, which is genetically and clinically correlated to schizophrenia (Bulik-Sullivan et al., 2015), does not show a similar increased diversity. We did observe a strong inverse correlation between increased diversity and estimated cell abundance of a population of T-cells in healthy controls. Even though this finding is based on indirect cell count measures using DNA methylation data (Horvath & Levine, 2015), the significant correlation highlights a likely close connection between the immune system and the blood microbiome, a relationship that has been documented before (Belkaid & Hand, 2014). More extensive cell count measures and/or better markers of immune sensing of microbial products could be used to study this relationship more directly. In the absence of a direct link with genetic susceptibility and the reported correlation with the immune system, we hypothesize that the observed effect in schizophrenia may be mostly a consequence of disease. This may be affected by lifestyle and/or health status differences of schizophrenia patients, including smoking, treatment plans, (chronic) infection, GI status, the use of probiotics, antibiotics and other drug use or other environmental exposures. Future targeted and/or longitudinal studies with larger sample sizes, detailed clinical phenotypes and more in-depth sequencing are needed to corroborate this hypothesis. Another interesting direction for future work is to study gut permeability in the context of our findings more directly. For example, how does damage to the gut (such as measured using I-FABP) affect observed microbial diversity in blood? These studies would likely result in an expanded understanding of the functional mechanisms underlying the connection between the human immune system, microbiome, and disease etiology. In particular, we hope that these future efforts will provide a useful quantitative and qualitative assessment of the microbiome and its role across the gut-blood barrier in the context of psychiatric disorders.

Secondly, we have developed a novel computational approach (ImReP) for reconstruction of Ig immune repertoires using RNA-Seq data. We demonstrate the ability of ImReP to efficiently extract Ig- derived reads from the RNA-Seq data and accurately assemble the corresponding hypervariable region sequence. The proposed algorithm can accurately assemble CDR3 sequences of Ig receptors despite the presence of sequencing errors and short read length. Simulations generated using various read lengths and coverage depth show that ImReP consistently outperforms existing methods in terms of precision and recall rates.

We have demonstrated the feasibility of applying RNA-Seq to study the adaptive immune repertoire. Although RNA-Seq lacks the sequencing depth of targeted sequencing (Rep-Seq), it can compensate by examining a larger sample size. Using ImReP, we have created the first systematic atlas of immunological sequence for Ig receptor repertories across diverse human tissues. This provides a rich resource for comparative analysis of a range of tissue types, most of which have not been studied before. The atlas of immune repertoires, available with the paper, is one the largest collection of CDR3 sequences and tissue types. We anticipate that this database will enhance future studies in areas such as immunology and contribute to the development of therapies for human diseases.

Using RNA-Seq to study immune repertoires has some advantages, including the ability to simultaneously capture clonotype populations from all the chains during a single run. It also allows simultaneous detection of overall transcriptional responses of the adaptive immune system, by comparing changes in the number of Ig transcripts to the much larger transcriptome. Given a large number of large-scale RNA-Seq datasets becoming available, we look forward to scaling up the atlas of immune receptors in order to provide valuable insights into immune responses across various autoimmune diseases, allergies, and cancers.

We have demonstrated that important biological and clinical information that are hidden in contemporary RNA Sequencing analysis in can be discovered by comprehensive re-analysis of RNA-Seq data. We believe that development in more analysis methods that investigate past the contemporary RNA-Sequencing analysis will further enrich the knowledge that researchers can illuminate from ever-growing RNA-sequencing datasets.

Acknowledgements We thank Loes M Olde Loohuis, Igor Mandric, Nicolas Strauli, Franziska Gruhl, Hagit T. Porath, Dennis Montoya, Jeremy Rotman, Kevin Hsieh, Linus Chen, Will Van Der Wey, Jiem R. Ronas, Timothy Daley, Stephanie Christenson, Benjamin Statz, Douglas Yao, Agata Wesolowska-Andersen, Roberto Spreafico, Cydney Rios, Celeste Eng, Andrew D. Smith, Ryan D. Hernandez, Roel A. Ophoff, Jose Rodriguez Santana, Erez Y. Levanon, Prescott G. Woodruff, Esteban Burchard, Max A. Seibold, Sagiv Shifman, Anil P.S. Ori, Guillaume Jospin, David Koslicki, Timothy Wu, Marco P. Boks, Catherine Lomen-Hoerth, Martina Wiedau-Pazos, Roberto Spreafico, K. Mark Ansel, Rita M. Cantor, Willem M. de Vos, René S. Kahn, Maura Rossetti, and Alex Zelikovsky for their help in this work.

Disclosure
The authors declare no competing interests.

References

Abu-Shanab, A., & Quigley, E. M. M. (2010). The role of the gut microbiota in nonalcoholic fatty liver disease. *Nature Reviews Gastroenterology & Hepatology, 7*(12), 691–701. https://doi.org/10.1038/nrgastro.2010.172.

Adler, C. J., Dobney, K., Weyrich, L. S., Kaidonis, J., Walker, A. W., Haak, W., ... Cooper, A. (2013). Sequencing ancient calcified dental plaque shows changes in oral microbiota with dietary shifts of the Neolithic and Industrial revolutions. *Nature genetics, 45*(4), 450–5, 455e1. https://doi.org/10.1038/ng.2536.

Amar, J., Serino, M., Lange, C., Chabo, C., Iacovoni, J., Mondot, S., ... Burcelin, R. (2011). Involvement of tissue bacteria in the onset of diabetes in humans: Evidence for a concept. *Diabetologia, 54*(12), 3055–3061. https://doi.org/10.1007/s00125-011-2329-8.

Anders, S., Pyl, P. T., & Huber, W. (2014). HTSeq–A Python framework to work with high-throughput sequencing data. *Bioinformatics*, btu638.

Andrews, S. (2010). FastQC: A quality control tool for high throughput sequence data.

Aryee, M. J., Jaffe, A. E., Corrada-Bravo, H., Ladd-Acosta, C., Feinberg, A. P., Hansen, K. D., et al. (2014). Minfi: A flexible and comprehensive bioconductor package for the analysis of Infinium DNA methylation microarrays. *Bioinformatics, 30*(10), 1363–1369. https://doi.org/10.1093/bioinformatics/btu049.

Baruzzo, G., Hayer, K. E., Kim, E. J., Di Camillo, B., FitzGerald, G. A., & Grant, G. R. (2016). Simulation-based comprehensive benchmarking of RNA-seq aligners. *Nature Methods, 14*(2), 135–139.

Bazak, L., Haviv, A., Barak, M., Jacob-Hirsch, J., Deng, P., Zhang, R., ... Levanon, E. Y. (2014). A-to-I RNA editing occurs at over a hundred million genomic sites, located in a majority of human genes. *Genome Research, 24*(3), 365–376.

Beck, J. M., Young, V. B., & Huffnagle, G. B. (2012). The microbiome of the lung. *Translational Research: The Journal of Laboratory and Clinical Medicine, 160*(4), 258–266. https://doi.org/10.1016/j.trsl.2012.02.005.

Belkaid, Y., & Hand, T. W. (2014). Role of the microbiota in immunity and inflammation. *Cell, 157*(1), 121–141. https://doi.org/10.1016/j.cell.2014.03.011.

Ben-Amor, K., Heilig, H., Smidt, H., Vaughan, E. E., Abee, T., & de Vos, W. M. (2005). Genetic diversity of viable, injured, and dead fecal bacteria assessed by fluorescence-activated cell sorting and 16S rRNA gene analysis. *Applied and Environmental Microbiology, 71*(8), 4679–4689.

Benichou, J., Ben-Hamo, R., Louzoun, Y., & Efroni, S. (2012). Rep-seq: Uncovering the immunological repertoire through next-generation sequencing. *Immunology, 135*(3), 183–191.

Blachly, J. S., Ruppert, A. S., Zhao, W., Long, S., Flynn, J., Flinn, I., … Rassenti, L. Z. (2015). Immunoglobulin transcript sequence and somatic hypermutation computation from unselected RNA-seq reads in chronic lymphocytic leukemia. *Proceedings of the National Academy of Sciences, 112*(14), 4322–4327. http://www.ncbi.nlm.nih.gov/pubmed/25787252 http://www.pubmedcentral.nih.gov/articlerender.fcgi?artid=PMC4394264 http://www.pnas.org/lookup/doi/10.1073/pnas.1503587112.

Bolotin, D. A., Poslavsky, S., Mitrophanov, I., Shugay, M., Mamedov, I. Z., Putintseva, E. V., et al. (2015). MiXCR: Software for comprehensive adaptive immunity profiling. *Nature Methods, 12*(5), 380–381.

Brown, S. D., Raeburn, L. A., & Holt, R. A. (2015). Profiling tissue-resident T cell repertoires by RNA sequencing. *Genome Medicine, 7*(1), 1–8.

Bulik-Sullivan, B., Finucane, H. K., Anttila, V., Gusev, A., … Day, F. R. (2015). An atlas of genetic correlations across human diseases and traits. *Nature Genetics, 47*(11), 1236–1241. https://doi.org/10.1038/ng.3406.

Camacho, C., Coulouris, G., Avagyan, V., Ma, N., Papadopoulos, J., Bealer, K., et al. (2009). BLAST+: Architecture and applications. *BMC Bioinformatics, 10*(1), 421. https://doi.org/10.1186/1471-2105-10-421.

Cayrou, C., Sambe, B., Armougom, F., Raoult, D., & Drancourt, M. (2013). Molecular diversity of the *planctomycetes* in the human gut microbiota in France and Senegal. *APMIS, 121*(11), 1082–1090. https://doi.org/10.1111/apm.12087.

Chen, Y., Lemire, M., Choufani, S., Butcher, D. T., Grafodatskaya, D., Zanke, B. W., … Weksberg, R. (2013). Discovery of cross-reactive probes and polymorphic CpGs in the Illumina Infinium HumanMethylation450 microarray. *Epigenetics, 8*(2), 203–209.

Cho, I., & Blaser, M. J. (2012). The human microbiome: At the interface of health and disease. *Nature Reviews Genetics, 13*(4), 260–270. https://doi.org/10.1038/nrg3182.

Chuang, T.-J., Wu, C.-S., Chen, C.-Y., Hung, L.-Y., Chiang, T.-W., & Yang, M.-Y. (2015). NCLscan: Accurate identification of non-co-linear transcripts (fusion, trans-splicing and circular RNA) with a good balance between sensitivity and precision. *Nucleic Acids Research*, gkv1013.

Cloonan, N., Forrest, A. R. R., Kolle, G., Gardiner, B. B. A., Faulkner, G. J., Brown, M. K., … Robertson, A. J. (2008). Stem cell transcriptome profiling via massive-scale mRNA sequencing. *Nature Methods, 5*(7), 613–619.

Criscione, S. W., Zhang, Y., Thompson, W., Sedivy, J. M., & Neretti, N. (2014). Transcriptional landscape of repetitive elements in normal and cancer human cells. *BMC Genomics, 15*(1), 583. https://doi.org/10.1186/1471-2164-15-583.

Crooks, G. E., Hon, G., Chandonia, J.-M., & Brenner, S. E. (2004). WebLogo: A sequence logo generator. *Genome Research, 14*(6), 1188–1190.

Croucher, N. J., & Thomson, N. R. (2010). Studying bacterial transcriptomes using RNA-seq. *Current Opinion in Microbiology, 13*(5), 619–624.

Darling, A. E., Jospin, G., Lowe, E., Matsen, F. A., Bik, H. M., & Eisen, J. A. (2014). PhyloSift: Phylogenetic analysis of genomes and metagenomes. *PeerJ, 2*, e243. https://doi.org/10.7717/peerj.243.

De Rossi, A., Roncella, S., Calabro, M. L., D'Andrea, E., Pasti, M., Panozzo, M., … Chieco-Bianchi, L. (1990). Infection of Epstein-Barr virus-transformed lymphoblastoid B cells by the human immunodeficiency virus: Evidence for a persistent and productive infection leading to B cell phenotypic changes. *European Journal of Immunology, 20*(9), 2041–2049.

de Vos, W. M., & de Vos, E. A. J. (2012). Role of the intestinal microbiome in health and disease: From correlation to causation. *Nutrition Reviews, 70,* S45–S56. https://doi.org/10.1111/j.1753-4887.2012.00505.x.

DeWitt, W. S., Lindau, P., Snyder, T. M., Sherwood, A. M., Vignali, M., Carlson, C. S., ... Robins, H. S. (2016). A public database of memory and naive B-cell receptor sequences. *PLoS One, 11*(8), e0160853.

Dickerson, F., Severance, E., & Yolken, R. (2017). The microbiome, immunity, and schizophrenia and bipolar disorder. *Brain, Behavior, and Immunity, 62,* 46–52. https://doi.org/10.1016/j.bbi.2016.12.010.

Dinan, T. G., Borre, Y. E., & Cryan, J. F. (2014). Genomics of schizophrenia: Time to consider the gut microbiome? *Molecular Psychiatry, 19*(12), 1252–1257. https://doi.org/10.1038/mp.2014.93.

Dobin, A., Davis, C. A., Schlesinger, F., Drenkow, J., Zaleski, C., Jha, S., ... Gingeras, T. R. (2013). STAR: Ultrafast universal RNA-seq aligner. *Bioinformatics, 29*(1), 15–21. https://doi.org/10.1093/bioinformatics/bts635.

Drennan, M. R. (1942). What is "Sterile blood"? *BMJ, 2*(4269), 526. https://doi.org/10.1136/bmj.2.4269.526.

El Kaoutari, A., Armougom, F., Gordon, J. I., Raoult, D., & Henrissat, B. (2013). The abundance and variety of carbohydrate-active enzymes in the human gut microbiota. *Nature Reviews Microbiology, 11*(7), 497–504. https://doi.org/10.1038/nrmicro3050.

Fellerhoff, B., Laumbacher, B., Mueller, N., Gu, S., & Wank, R. (2006). Associations between *Chlamydophila* infections, schizophrenia and risk of HLA-A10. *Molecular Psychiatry, 12*(3), 264–272. https://doi.org/10.1038/sj.mp.4001925.

Foster, J. A., & Neufeld, K.-A. M. (2013). Gut-brain axis: How the microbiome influences anxiety and depression. *Trends in Neurosciences, 36*(5), 305–312. https://doi.org/10.1016/j.tins.2013.01.005.

Freeman, J. D., Warren, R. L., Webb, J. R., Nelson, B. H., & Holt, R. A. (2009). Profiling the T-cell receptor beta-chain repertoire by massively parallel sequencing. *Genome Research, 19*(10), 1817–1824.

Georgiou, G., Ippolito, G. C., Beausang, J., Busse, C. E., Wardemann, H., & Quake, S. R. (2014). The promise and challenge of high-throughput sequencing of the antibody repertoire. *Nature Biotechnology, 32*(2), 158–168.

Grabherr, M. G., Haas, B. J., Yassour, M., Levin, J. Z., Thompson, D. A., Amit, I., ... Regev, A. (2011). Full-length transcriptome assembly from RNA-Seq data without a reference genome. *Nature Biotechnology, 29*(7), 644–52. https://doi.org/10.1038/nbt.1883.

Greenblum, S., Turnbaugh, P. J., & Borenstein, E. (2012). Metagenomic systems biology of the human gut microbiome reveals topological shifts associated with obesity and inflammatory bowel disease. *Proceedings of the National Academy of Sciences, 109*(2), 594–599.

GTEx Consortium. (2015). The genotype-tissue expression (GTEx) pilot analysis: Multitissue gene regulation in humans. *Science, 348*(6235), 648–660.

GTEx Consortium, Aguet, F., Brown, A. A., Castel, S. E., Davis, J. R., He, Y., ... Zhu, J. (2017). Genetic effects on gene expression across human tissues. *Nature, 550*(7675), 204–213. https://doi.org/10.1038/nature24277.

Hoi, K. H., & Ippolito, G. C. (2013). Intrinsic bias and public rearrangements in the human immunoglobulin Vλ light chain repertoire. *Genes and Immunity, 14*(4), 271–276.

Horvath, S. (2013). DNA methylation age of human tissues and cell types. *Genome Biology, 14*(10), R115. https://doi.org/10.1186/gb-2013-14-10-r115.

Horvath, S., & Levine, A. J. (2015). HIV-1 infection accelerates age according to the epigenetic clock. *Journal of Infectious Diseases, 212*(10), 1563–1573. https://doi.org/10.1093/infdis/jiv277.

Hou, S., Makarova, K. S., Saw, J. H. W., Senin, P., Ly, B. V, Zhou, Z., ... Alam, M. (2008). Complete genome sequence of the extremely acidophilic methanotroph isolate V4, Methylacidiphilum infernorum, a representative of the bacterial phylum Verrucomicrobia. *Biology Direct, 3*(1), 26. https://doi.org/10.1186/1745-6150-3-26.

Houseman, E., Accomando, W. P., Koestler, D. C., Christensen, B. C., Marsit, C. J., Nelson, H. H., … Kelsey, K. T. (2012). DNA methylation arrays as surrogate measures of cell mixture distribution. *BMC Bioinformatics, 13*(1), 86. https://doi.org/10.1186/1471-2105-13-86.

Human Microbiome Project. (2012). Structure, function and diversity of the healthy human microbiome. *Nature, 486*(7402), 207–214. https://doi.org/10.1038/nature11234.

Humphrys, M. S., Creasy, T., Sun, Y., Shetty, A. C., Chibucos, M. C., Drabek, E. F., … & Shou, H. (2013). Simultaneous transcriptional profiling of bacteria and their host cells. *PloS One, 8*(12), e80597.

Inman, C. F., Murray, T. Z., Bailey, M., & Cose, S. (2012). Most B cells in non-lymphoid tissues are naïve. *Immunology and Cell Biology, 90*(2), 235–242. https://doi.org/10.1038/icb.2011.35.

Jeck, W. R., & Sharpless, N. E. (2014). Detecting and characterizing circular RNAs. *Nature Biotechnology, 32*(5), 453–461. https://doi.org/10.1038/nbt.2890.

Jin, Y., Tam, O. H., Paniagua, E., & Hammell, M. (2015). TEtranscripts: A package for including transposable elements in differential expression analysis of RNA-seq datasets. *Bioinformatics*, btv422.

Jorth, P., Turner, K. H., Gumus, P., Nizam, N., Buduneli, N., & Whiteley, M. (2014). Metatranscriptomics of the human oral microbiome during health and disease. *MBio, 5*(2). https://doi.org/10.1128/mbio.01012-14.

Jost, L. (2007). Partitioning diversity into independent alpha and beta components. *Ecology, 88*(10), 2427–2439. https://doi.org/10.1890/06-1736.1.

Kim, D., Pertea, G., Trapnell, C., Pimentel, H., Kelley, R., & Salzberg, S. L. (2013). TopHat2: Accurate alignment of transcriptomes in the presence of insertions, deletions and gene fusions. *Genome Biology, 14*(4), R36. https://doi.org/10.1186/gb-2013-14-4-r36.

Kim, D., & Salzberg, S. L. (2011). TopHat-Fusion: An algorithm for discovery of novel fusion transcripts. *Genome Biology, 12*(8), R72–R72. https://doi.org/10.1186/gb-2011-12-8-r72.

Koch, S., Larbi, A., Derhovanessian, E., Özcelik, D., Naumova, E., & Pawelec, G. (2008). Multiparameter flow cytometric analysis of CD4 and CD8 T cell subsets in young and old people. *Immunity & Ageing, 5*(1), 6. https://doi.org/10.1186/1742-4933-5-6.

Koleff, P., Gaston, K. J., & Lennon, J. J. (2003). Measuring beta diversity for presence-absence data. *Journal of Animal Ecology, 72*(3), 367–382. https://doi.org/10.1046/j.1365-2656.2003.00710.x.

Koren, O., Spor, A., Felin, J., Fak, F., Stombaugh, J., Tremaroli, V., … Backhed, F. (2010). Human oral, gut, and plaque microbiota in patients with atherosclerosis. *Proceedings of the National Academy of Sciences, 108*(Supplement_1), 4592–4598. https://doi.org/10.1073/pnas.1011383107.

Kostic, A. D., Ojesina, A. I., Pedamallu, C. S., Jung, J., Verhaak, R. G. W., Getz, G., et al. (2011). PathSeq: Software to identify or discover microbes by deep sequencing of human tissue. *Nature Biotechnology, 29*(5), 393–396.

Kuchenbecker, L., Nienen, M., Hecht, J., Neumann, A. U., Babel, N., Reinert, K., et al. (2015). IMSEQ—A fast and error aware approach to immunogenetic sequence analysis. *Bioinformatics, 31*(18), 2963–2971. https://doi.org/10.1093/bioinformatics/btv309.

Lagier, J.-C., Million, M., Hugon, P., Armougom, F., & Raoult, D. (2012). Human gut microbiota: Repertoire and variations. *Frontiers in Cellular and Infection Microbiology, 2*. https://doi.org/10.3389/fcimb.2012.00136.

Landsverk, O. J. B., Snir, O., Casado, R. B., Richter, L., Mold, J. E., Réu, P., … Øyen, O. M. (2017). Antibody-secreting plasma cells persist for decades in human intestine. *Journal of Experimental Medicine*, jem—20161590.

Lefranc, M.-P., Giudicelli, V., Duroux, P., Jabado-Michaloud, J., Folch, G., Aouinti, S., … Kossida, S. (2015). IMGT®, the international ImMunoGeneTics information system® 25 years on. *Nucleic Acids Res., 43*(Database issue), D413–D422.

Li, B., Li, T., Pignon, J.-C., Wang, B., Wang, J., Shukla, S. A., … Wu, C. (2016). Landscape of tumor-infiltrating T cell repertoire of human cancers. *Nature Genetics, 48*(7), 725–732.

Li, S., Tighe, S. W., Nicolet, C. M., Grove, D., Levy, S., Farmerie, W., … Kim, D. (2014). Multi-platform assessment of transcriptome profiling using RNA-seq in the ABRF next-generation sequencing study, *Nature biotechnology, 32*(9), 915–925. https://doi.org/10.1038/nbt.2972.

Lombardo, K. A., Coffey, D. G., Morales, A. J., Carlson, C. S., Towlerton, A. M. H., Gerdts, S. E., … Warren, E. H. (2017). High-throughput sequencing of the B-cell receptor in African Burkitt lymphoma reveals clues to pathogenesis. *Blood Advances, 1*(9), 535–544.

Lopez, D., Montoya, D., Ambrose, M., Lam, L., Briscoe, L., Adams, C., … Pellegrini, M. (2017). SaVanT: A web-based tool for the sample-level visualization of molecular signatures in gene expression profiles. *BMC Genomics, 18*(1).

McClure, R., Balasubramanian, D., Sun, Y., Bobrovskyy, M., Sumby, P., Genco, C. A., … Tjaden, B. (2013). Computational analysis of bacterial RNA-Seq data. *Nucleic Acids Research, 41*(14), e140–e140. https://doi.org/10.1093/nar/gkt444.

McLaughlin, R. W., Vali, H., Lau, P. C. K., Palfree, R. G. E., Ciccio, A. De, Sirois, M., … Chan, E. C. S. (2002). Are there naturally occurring pleomorphic bacteria in the blood of healthy humans? *Journal of Clinical Microbiology, 40*(12), 4771–4775. https://doi.org/10.1128/jcm.40.12.4771-4775.2002.

Mihaela Pertea, J. T. M. S. L. S. (2015). StringTie enables improved reconstruction of a transcriptome from RNA-seq reads. *Nature Biotechnology, 33,* 290–295. https://doi.org/10.1038/nbt.3122.

Mose, L. E., Selitsky, S. R., Bixby, L. M., Marron, D. L., Iglesia, M. D., Serody, J. S., … Parker, J. S. (2016). Assembly-based inference of B-cell receptor repertoires from short read RNA sequencing data with V'DJer. *Bioinformatics,* btw526.

Nicolae, M., Mangul, S., Mandoiu, I. I., & Zelikovsky, A. (2011). Estimation of alternative splicing isoform frequencies from RNA-Seq data. *Algorithms for Molecular Biology, 6*(1), 9.

Nikkari, S., McLaughlin, I. J., Bi, W., Dodge, D. E., & Relman, D. A. (2001). Does blood of healthy subjects contain bacterial ribosomal DNA? *Journal of Clinical Microbiology, 39*(5), 1956–1959. https://doi.org/10.1128/jcm.39.5.1956-1959.2001.

Ozsolak, F., & Milos, P. M. (2011). RNA sequencing: Advances, challenges and opportunities. *Nature Reviews Genetics, 12*(2), 87–98. https://doi.org/10.1038/nrg2934.

Païssé, S., Valle, C., Servant, F., Courtney, M., Burcelin, R., Amar, J., et al. (2016). Comprehensive description of blood microbiome from healthy donors assessed by 16S targeted metagenomic sequencing. *Transfusion, 56*(5), 1138–1147. https://doi.org/10.1111/trf.13477.

Perucheon, S., Chaoul, N., Burelout, C., Delache, B., Brochard, P., Laurent, P., … Richard, Y. (2009). Tissue-specific B-cell dysfunction and generalized memory B-cell loss during acute SIV infection. *PLoS ONE, 4*(6), e5966. https://doi.org/10.1371/journal.pone.0005966.

Philibert, P., Stoessel, A., Wang, W., Sibler, A.-P., Bec, N., Larroque, C., … Martineau, P. (2007). A focused antibody library for selecting scFvs expressed at high levels in the cytoplasm. *BMC Biotechnology, 7*(1), 1.

Poole, A., Urbanek, C., Eng, C., Schageman, J., Jacobson, S., O'Connor, B. P., … Lutz, S. (2014). Dissecting childhood asthma with nasal transcriptomics distinguishes subphenotypes of disease. *Journal of Allergy and Clinical Immunology, 133*(3), 670–678.

Porath, H. T., Carmi, S., & Levanon, E. Y. (2014). A genome-wide map of hyper-edited RNA reveals numerous new sites. *Nature Communications, 5,* 4726. https://doi.org/10.1038/ncomms5726.

Potgieter, M., Bester, J., Kell, D. B., & Pretorius, E. (2015). The dormant blood microbiome in chronic, inflammatory diseases. *FEMS Microbiology Reviews, 39*(4), 567–591. https://doi.org/10.1093/femsre/fuv013.

Price, M. E., Cotton, A. M., Lam, L. L., Farre, P., Emberly, E., Brown, C. J., … Kobor, M. S. (2013). Additional annotation enhances potential for biologically-relevant analysis of the Illumina Infinium HumanMethylation450 BeadChip array. *Epigenetics & Chromatin, 6*(1), 4. https://doi.org/10.1186/1756-8935-6-4.

Probst, A. J., Auerbach, A. K., & Moissl-Eichinger, C. (2013). Archaea on human skin. *PloS One, 8*(6), e65388. https://doi.org/10.1371/journal.pone.0065388.

Putintseva, E. V, Britanova, O. V, Staroverov, D. B., Merzlyak, E. M., Turchaninova, M. A., Shugay, M., ... Chudakov, D. M. (2013). Mother and child T cell receptor repertoires: Deep profiling study. *Frontiers in Immunology, 4*, 463. https://doi.org/10.3389/fimmu.2013.00463.

Rajewsky, K., Forster, I., & Cumano, A. (1987). Evolutionary and somatic selection of the antibody repertoire in the mouse. *Science, 238*(4830), 1088–1094.

Ripke, S., O'Dushlaine, C., Chambert, K., Moran, J. L., Kahler, A. K., Akterin, S., ... Sullivan, P. F. (2013). Genome-wide association analysis identifies 13 new risk loci for schizophrenia. *Nature Genetics, 45*(10), 1150–1159. https://doi.org/10.1038/ng.2742.

Salter, S. J., Cox, M. J., Turek, E. M., Calus, S. T., Cookson, W. O., Moffatt, M. F., ... Walker, A. W. (2014). Reagent and laboratory contamination can critically impact sequence-based microbiome analyses. *BMC Biology, 12*(1), 87.

Sato, J., Kanazawa, A., Ikeda, F., Yoshihara, T., Goto, H., Abe, H., ... Watada, H. (2014). Gut dysbiosis and detection of "Live gut bacteria" in blood of Japanese patients with type 2 diabetes. *Diabetes Care, 37*(8), 2343–2350. https://doi.org/10.2337/dc13-2817.

Schizophrenia Working Group of the Psychiatric Genomics Consortium. (2014). Biological insights from 108 schizophrenia-associated genetic loci. *Nature, 511*(7510), 421–427. https://doi.org/10. 1038/nature13595.

Segata, N., Waldron, L., Ballarini, A., Narasimhan, V., Jousson, O., & Huttenhower, C. (2012). Metagenomic microbial community profiling using unique clade-specific marker genes. *Nature Methods, 9*(8), 811–814. https://doi.org/10.1038/nmeth.2066.

Seqc/Maqc-Iii Consortium. (2014). A comprehensive assessment of RNA-seq accuracy, reproducibility and information content by the Sequencing Quality Control Consortium. *Nature Biotechnology, 32*(9), 903–914. https://doi.org/10.1038/nbt.2957.

Simpson, E. H. (1949). Measurement of diversity. *Nature, 163*(4148), 688. https://doi.org/10.1038/ 163688a0.

Siragusa, E., Weese, D., & Reinert, K. (2013). Fast and accurate read mapping with approximate seeds and multiple backtracking. *Nucleic Acids Research, 41*(7), e78–e78.

Spadoni, I., Zagato, E., Bertocchi, A., Paolinelli, R., Hot, E., Di Sabatino, A., ... Rescigno, M. (2015). A gut-vascular barrier controls the systemic dissemination of bacteria. *Science, 350*(6262), 830–834. https://doi.org/10.1126/science.aad0135.

Spreafico, R., Rossetti, M., van Loosdregt, J., Wallace, C. A., Massa, M., Magni-Manzoni, S., ... Albani, S. (2016). A circulating reservoir of pathogenic-like CD4+ T cells shares a genetic and phenotypic signature with the inflamed synovial micro-environment. *Annals of the Rheumatic Diseases, 75*(2), 459–465.

Strauli, N. B., & Hernandez, R. D. (2016). Statistical inference of a convergent antibody repertoire response to influenza vaccine. *Genome Medicine, 8*(1), 1. article.

Strauli, N., & Hernandez, R. (2015). Statistical inference of a convergent antibody repertoire response to influenza vaccine. *BioRxiv*, 25098.

Strong, M. J., Xu, G., Morici, L., Bon-Durant, S. S., Baddoo, M., Lin, Z., ... Flemington, E. K. (2014). Microbial Contamination in next generation sequencing: Implications for sequence-based analysis of clinical samples. *PLoS Pathogens, 10*(11), e1004437. https://doi.org/10.1371/journal. ppat.1004437.

Stubbington, M. J. T., Lönnberg, T., Proserpio, V., Clare, S., Speak, A. O., Dougan, G., & Teichmann, S. A. (2016). T cell fate and clonality inference from single-cell transcriptomes. *Nature Methods, 13*(4), 329–332.

Sultan, M., Amstislavskiy, V., Risch, T., Schuette, M., Dökel, S., Ralser, M., ... Yaspo, M.-L. (2014). Influence of RNA extraction methods and library selection schemes on RNA-seq data. *BMC Genomics, 15*(1), 675.

Tang, F., Barbacioru, C., Wang, Y., Nordman, E., Lee, C., Xu, N., ... Lao, K. (2009). mRNA-Seq whole-transcriptome analysis of a single cell. *Nature Methods, 6*(5), 377–382.

Tarailo-Graovac, M., & Chen, N. (2009). Using RepeatMasker to identify repetitive elements in genomic sequences. *Current Protocols in Bioinformatics*, 4–10.

Trapnell, C., Williams, B. a, Pertea, G., Mortazavi, A., Kwan, G., van Baren, M. J., ... Pachter, L. (2010). Transcript assembly and quantification by RNA-Seq reveals unannotated transcripts and isoform switching during cell differentiation. *Nature Biotechnology, 28*(5), 511–515. https://doi.org/10.1038/nbt.1621.

Truong, D. T., Franzosa, E. A., Tickle, T. L., Scholz, M., Weingart, G., Pasolli, E., ... Segata, N. (2015). MetaPhlAn2 for enhanced metagenomic taxonomic profiling. *Nature Methods, 12*(10), 902–903.

Turnbaugh, P. J., Hamady, M., Yatsunenko, T., Cantarel, B. L., Duncan, A., Ley, R. E., ... Gordon, J. I. (2008). A core gut microbiome in obese and lean twins. *Nature, 457*(7228), 480–484. https://doi.org/10.1038/nature07540.

Turnbaugh, P. J., Ley, R. E., Mahowald, M. A., Magrini, V., Mardis, E. R., & Gordon, J. I. (2006). An obesity-associated gut microbiome with increased capacity for energy harvest. *Nature, 444*(7122), 1027–1031. https://doi.org/10.1038/nature05414.

Wang, X.-S., Prensner, J. R., Chen, G., Cao, Q., Han, B., Dhanasekaran, S. M., ... Chinnaiyan, A. M. (2009). An integrative approach to reveal driver gene fusions from paired-end sequencing data in cancer. *Nature Biotechnology, 27*(11), 1005–11. https://doi.org/10.1038/nbt.1584.

Warren, R. L., Freeman, J. D., Zeng, T., Choe, G., Munro, S., Moore, R., ... Holt, R. A. (2011). Exhaustive T-cell repertoire sequencing of human peripheral blood samples reveals signatures of antigen selection and a directly measured repertoire size of at least 1 million clonotypes. *Genome Research, 21*(5), 790–797.

Warren, R. L., Nelson, B. H., & Holt, R. A. (2009). Profiling model T-cell metagenomes with short reads. *Bioinformatics, 25*(4), 458–464.

Whittaker, R. H. (1972). Evolution and measurement of species diversity. *Taxon, 21*(2/3), 213. https://doi.org/10.2307/1218190.

Wu, C.-S., Yu, C.-Y., Chuang, C.-Y., Hsiao, M., Kao, C.-F., Kuo, H.-C., et al. (2014). Integrative transcriptome sequencing identifies trans-splicing events with important roles in human embryonic stem cell pluripotency. *Genome Research, 24*(1), 25–36.

Wu, S., Yi, J., Zhang, Y., Zhou, J., & Sun, J. (2015). Leaky intestine and impaired microbiome in an amyotrophic lateral sclerosis mouse model. *Physiological Reports, 3*(4), e12356. https://doi.org/10.14814/phy2.12356.

Yan, M., Pamp, S. J., Fukuyama, J., Hwang, P. H., Cho, D. Y., Holmes, S., & Relman, D. A. (2013). Nasal microenvironments and interspecific interactions influence nasal microbiota complexity and *S. aureus* carriage. *Cell Host and Microbe, 14*(6), 631–640. https://doi.org/10.1016/j.chom.2013.11.005.

Ye, J., Ma, N., Madden, T. L., & Ostell, J. M. (2013). IgBLAST: An immunoglobulin variable domain sequence analysis tool. *Nucleic Acids Research,* gkt382.

Yu, H.-P., Chiu, Y.-W., Lin, H.-H., Chang, T.-C., & Shen, Y.-Z. (1991). Blood content in guinea-pig tissues: Correction for the study of drug tissue distribution. *Pharmacological Research, 23*(4), 337–347.

Zhang, X.-O., Dong, R., Zhang, Y., Zhang, J.-L., Luo, Z., Zhang, J., ... Yang, L. (2016). Diverse alternative back-splicing and alternative splicing landscape of circular RNAs. *Genome Research.* https://doi.org/10.1101/gr.202895.115.